Let the Sun Rise for Extensive Solar Power Generation in India.

A Practical Guide for Total Engineering of MW capacity Solar PV Power Project

A.S.Kapur

DEDICATION

This book is dedicated to one of the India's best Solar Power EPC company-

M/S Rays Power Infra, a Jaipur based company and its young Directors Mr.Ketan Mehta and Mr.Pawan Sharma

Rays power infra, the pioneers in the construction of solar power plants, has to its credit, construction of above 200 MW capacity plants, with present maximum size of a single plant as 50 MW.
Theirs, this achievement, is accomplished in a very short period of about three years.
Their designed and constructed PV solar power plants are achieving extremely high PLANT PERFORMANCE RATIO and very high Capacity Utilisation Factor.
They are also in the field of construction of 220kv, 132kv and 66 kv grid sub stations and the associated transmission lines.
The author also dedicates this write up to the young team of engineers working in this company. To name a few of them- Narendra, Rahul, Akhil, Inderjeet, Manish, Vijay, Anjan, Harish, Vinit and Himanshu. The author enjoys working with these young persons and learns many technical intricacies of the complicated solar power engineering.

All are also welcome to point out mistakes and make corrections in the technical aspects of the solar power systems.

Thanks and Acknowledgements for References

"This book is inspired by the GUIDE by IFC, which is a theoretical and techno-commercial write up for Developers and Investors of India, who want to go in to the solar power generation field.Thanks to IFC for this inspiration."

1.0 The actual field experience of the author, which is put in into this book, for the works carried out under his supervision at many sites in Rajasthan, Andhra and Karnataka.

1.1 The theory, engineering and concepts of PV power generation system, followed world over and taught in the classes. Also-;
Standard Technical practices, as per the text books and feedbacks, have been accommodated in this book on purely technical reasons.

1.2 Germany is the leading solar power generating country in the world, which has constructed solar power projects of capacity 36000 MW till now. A majority of Solar Power projects world over are based on German technology, widely prevalent as a de-factor PRACTICAL across the world. At some places the description resembles and has a technology references to it converted to Indian Methodology.

1.3 Italy has constructed 17000 MW capacity solar power projects so far. Italian engineers have shared the technical aspects of their projects with the world for wider implementation and technology sharing in the benefit of green technology initiatives. At some places the description resembles and has a flavor of Italian themed implementation, which is purely un-intentional.

1.4 Inverter and module manufacturers' guidelines as published by them, time and again in various formats of information sharing.

1.5 Seminars and workshops conducted by Gujarat Energy Research and Management Institute.

1.6 A text book on Large Scale Solar Power System Design-; an engineering guide for grid connected solar power generation by Peter Gevorkian. This book is ultimate in PV Solar Power concept, but its concept has to be modified for making it workable for the Indian power systems.

1.7 International Conference on energy systems and feedbacks.

1.8 International guidelines for the certifications of photovoltaic system components and grid connected systems of IEA.

1.9 Wikipedia-the free encyclopedia.

1.10 Indian Electricity Act guidelines/regulations/norms.

1.11 The state Discom and Transco rules and regulations.

Finally-;

THIS WRITE UP IS A COLLECTION OF THE BEST THEORIES, PROPOGATED BY THE WORLD'S EXPERTS, IN THE FIELD OF PV SOLAR POWER GENERATION. THEIR THEORIES ARE SUPREMPOSED WITH THE PRACTICAL FIELD EXPERIENCE OF THE AUTHOR. AUTHOR ACKNOWLEDGES THEIR CONTRIBUTION TO THE WELFARE OF ENTIRE MANKIND AND OFFERS THANKS.

Preface

The PV SOLAR POWER PLANT ENGINEERING is a developing exercise. It is -;

- Technically driven,
- Commercially driven with cost reducing considerations
- Quality driven
- Maximum energy output with minimum input driven

For this reason, it is necessary to have a team of experienced solar professionals and engineers, with field-proven expertise in all engineering fields, relevant to solar PV power applications. This team will design, build, commission, operate and maintain highly efficient solar power plants from start to completion, that too within desired specifications, budget and time lines.

Procurement activity, with leading equipment suppliers, to procure suitable equipments for solar power plants, and also site works with contractors to deliver clean, and cost-effective energy to the grid, is an activity which requires total know how of the complete system.

Designers, all over the world, have made the PV solar Power generation Plant concept almost standardized. But there is a need to improve the system economically and technically by adding the latest technical advancements available in the total plant concept. The system covers modules, module mounting structures, power pooling network, power carrying network, inverters, transformers, protection gears of 33/66/132/220 kv breakers and power evacuation feeders. The govt. desired metering is also standardized.

Also, the reduction in the overall cost of the new plants, and ease in the construction and O&M is the main issue, which the designers face in every new design. Therefore it is absolutely necessary, that-;

We stick to the fundamentals of electrical engineering,

We reduce the quantity of the items required to complete the project,

We reduce the civil construction cost,

We make the O&M simpler;

We give special care to the weakest link---cables.

We reduce the chances of faults to bare minimum

We utilize the sun to maximum day hour's window, especially in winter.

Based on above facts and findings, this book contains the detailed practical concept and guidelines for designing, constructing, and to operate and maintain a MW capacity, grid connected, PV solar power plant.

A.S.Kapur
Ex Dy. Chief Engineer
Rajasthan Vidyut Utpadan Nigam,Rajasthan- India

Head of Design & Engineering
Rays Power Infra, Jaipur, Rajasthan, India

+91-09928071814
Email: askapur@ymail.com

INDEX

INTRODUCTION

THE SUN RELEASES ENERGY, FREELY, FOR THE MANKIND TO HARNESS.

From sun we get heat and light. Both the energies are being harnessed for human use all over the world. We are talking about the light energy, which is converted to electrical energy using Photovoltaic Cells.

We have been taught that Solar radiation is ultraviolet, visible, infrared-energy, in the wavelength, ranges approximately, from 300 to 3000 nanometers (nm, 10^{-9} m) and has three components as detailed below-;

The global horizontal irradiance (**GHI**) falling onto the Earth's surface consists of-;

- The diffused horizontal irradiance (**DHI**) from the sky. DHI is measured by a pyranometer shaded from the direct sun.
- The direct normal irradiance (**DNI**) from the sun. And DNI is measured by a pyrheliometer with a narrow view that only measures the beam directly from the sun.
- The relation between the three is-; **GHI = DHI + DNI*cos(θ)**Where **θ** is the solar zenith angle (vertically above the location is 0°, horizontal is 90°). GHI is measured by a pyranometer with a hemispherical view, mounted horizontally.

This Radiant Energy from sun, which travels to earth, is in the form of waves. We call it Electromagnetic Radiation (EM). EM is a stream of photons which carry the energy. When this wave of energy strikes a Photovoltaic Cell, it is converted into Electricity.

Because of earths tilt on its axis, its elliptical orbit around Sun and associated location of the sun between tropic of Cancer and tropic of Capricorn during a year, the amount of radiant energy falling on earth varies accordingly.

Also sunrise to sun set movement of sun location in the sky results in the change of angle of incidence of the EM on the solar modules. The maximum energy absorption by solar modules takes place when it faces the sun directly/ perpendicularly. Therefore, because of above detailed variations, the designer's task becomes very complicated to design a system, so as to exploit, absorb and also convert the sun energy into electrical energy to its maximum level. Both the North Pole and South Pole are freezed, because the sun rays always reach there at 20 deg. angle with ground. Further, it is an established scientifically explained theory that-;

47% of solar energy reaching to earth gets converted to heat
20% of photons pass through without releasing power
2% of energy is lost in the system
31% of energy is converted into electricity.

This 31% of sun's gifted free energy is being harnessed by the PV Solar power plants world over. Needless to mention that present sources of energy are in the stage of extinction-coal is in low supply, oil is dwindling, water is scarce and atomic power is full of accidents/risks/difficult to manage. Therefore the nature has provided the renewable energy sources like Wind and Solar. This book deals into the intricacies of the solar power generating plants, their design, construction, operation and maintenance. **With this introduction we proceed as below to analyze every aspect of a PV Solar Power Project.**

A.S.KAPUR

The PV module is basically a Current Generator. The sun light releases the charge from the cell cathode, which travels to the cell Anode, thereby making a current to flow in the circuit. The voltage (about 0.5volt.) is automatically generated across the cell because of the charge flow. I am not going into the electronic theory of P-N-P and N-P-N junctions in this book, but engineers should study these theories to understand the working of a PV module.

A number of mini solar cells (70 to 90), connected in series, make a module. A number of solar modules (20or22or24), electrically connected in series, make up a PV generator, which generates electrical power (open circuit voltage, and resultant current as and when the circuit is completed), phenomenon being dependent on solar insolation on the module and the ambient temperature. This power (D.C.) is collected in string combiner boxes (SCB) by parallel connection of a number of strings. This consolidated power is then fed to inverter to invert it to give A.C. power.

The following parameters are the fundamental considerations for arriving at the final design of a PV solar power plant-;

-Annual kwh output per MW,

-Cost per MW of the total project

-Life of the project based on the quality of the main equipment's.

-Ease of operation and maintenance.

1.0.0 DESIGNING

In Indian solar scene the main aspect of the designing a solar power plant, is the COST of the Plant per Mw. There are many items where the designer can reduce the cost. Some of them are-;

- **i)** Land used per MW- Designers have achieved this value as 6/7 Bigha (167X167ft.) per MW. Or 3 to 4 acres.
- **ii)** To reduce the land consumption they have taken double/triple portrait placement of solar modules with row to row gap as 3.7 meters and a single pathway from main gate to IRs. Full utilization and no wastage of even an inch of the land is being targeted.
- **iii)** The quantity of 4 sq.mm. DC cable for string formation is reduced to 8/9 km per MW.
- **iv)** Strings are made with vertical configuration to reduce cable lengths.
- **v)** The seasonal tilt angles are taken as 5—13—26--44 degrees.
- **vi)** A CUF of 19%, PR of 79% and annual generation level of 16, 50,000 units per anum per MW or more can be assured, taking the seasonal tilts.
- **vii)** The module support structure weight is reduced to- 42 tons/MW.
- **viii)** The cost of the following three items is being bargained from the suppliers to get the best, but at the cheapest cost- SOLAR MODULES, INVERTERS (Rs.2.9 per Watt), Module Mounting Structures (63 Rs.per MW) and TRANSFORMERS.
- **ix)** Over all plant earthing and lightening protection cost can be reduced by utilizing the steel support structures, which are embedded in the earth, as earth electrodes. Since these supports are so many in numbers, they create as much parallel paths to earth, finally reducing the resultant resistance value of earthing system to less than 1-ohm.
- **x)** Earthing in DC Area has to be equipment specific and not a big mesh of earthing strip/rod, laid in the entire area.

xi) With above considerations the cost per MW has to be reduced to Rs.50000000.00 or even still less and the cost per unit of generation has to come down to Rs.5.25.(year 2014-December)

x). The following points are the main guidelines to design a PV solar power project.

- Selection of land/location, with availability of sun, and level/ land contour.
- Assessment of tilt angles, based on season and daily sun movement to get maximum output.
- Selection of main equipments such as inverter , PV module, transformer and circuit breaker
- Quality assurance plan for Procurement and works.
- Drawings of all possible arrangements of the total system from module to Grid.
- Calculation of the irradiation on the plane of the panels and study of the PV Syst and Meteonorm based assessment of power generation..
- Designing the cost effective and low loss system of inter connections between modules to strings to inverters to transformer to meters to power evacuation.
- Planning of AC System with protections, metering and power evacuation feeders.
- The technical documentation of the project.

1.0.1 The"PV Syst" software is the foremost tool, being used world over by every designer to calculate the total month wise expected generation and anticipated losses throughout the year. Therefore the knowledge of PV Syst is absolutely necessary for the design team.

Similarly the **"Meteonorm"** software is also an important tool to design the solar power projects. With these software's, we can check as to what will be Annual energy output from any capacity plant at the geographical location of the project. Even a day to day target can be fixed for level of generation of the plant based on the sun availability and atmospheric conditions, which this software automatically implants into the assessment.

Actually the designing work of the PV Solar plant starts from the analysis generated by these two tools. The moment you decide to install a plant at any location, you have to take the longitude and latitude of the location as a first step. Using the longitude and latitude of the selected location, you find out from these two tools as to what will be your daily and monthly generation level at the selected location and what will be the loss factor. The report generated by these two tools will advice you whether it is worth installing your project at this selected location. People say that in India the Rajasthan desert is the best location for PV plant installation. But these two tools may indicate that LEH in Kashmir area is a better location for a PV plant.
We have discussed these tools in chapters ahead.

Solar Farms Concept and Description...AIM-

- ➢ To analyze the environmental and engineering issues, for building solar plants, in the form of an energy park, where a number of small power plants are erected.
- ➢ Every plant in the solar farm jointly shares the common facility, mainly the power evacuation network, apart from other civil, O&M, security, SCADA and other facilities.
- ➢ One agency can be employed to design, construct, commission, operate and maintain all the small plants in the park.
- ➢ Technical knowhow of this one agency too will be available to the non-technical investor in solar field.

- This one agency can arrange land, settle the right of the way issues for power evacuation and arrange the liasoning with govt. agencies for all approvals starting from registration to commissioning.
- Common metering system and recovery of the revenue too can be done by this one agency.

1.0.2 THE FOLLOWING ACTIVITIES HAVE BEEN IDENTIFIED FOR THE DESIGNERS TO WORK ON-;

Detailed below is the list of all the areas on which the engineers have to work to achieve the goal of construction of the best solar PV power plant-'

Activity-1: Arrange on-site photovoltaic project assessment, inspection and study by experts-

Activity -2: Regulatory issues based on solar policy of the government.

Activity -3: Site engineering assessment

Activity-4: Plan PV system size, conceptual design elements, and specifications

Activity-5General Technical Specifications of main equipments.

Activity-6Arrange Quality Assurance Plan from start of the work on paper till completion of the plant.

Activity- 7Develop installed cost estimates for the solar farm

Activity-8; Explanation of special issues-PID, MPPT, Solar cable, Master slave configuration

Activity -9: Works planning and Execution

Activity-10; Conduct economic analysis of the proposed solar farm's characteristics

Activity-11; Prepare a final report and all drawings-SLD, Cabling, String formation and civil drawings

Activity -12; Commissioning of plant.

Activity -13; O&M of the plant.

Activity 1: Arrange on-site photovoltaic project assessment, inspection and study by experts- A detailed site visit report is to be prepared by experts; this report is briefly summarized below.

1.1 Site Visit Report- A note on the site visit is required indicating the following aspects-;

Approach road for heavy vehicles

Drinking water and sufficient quantity of water for construction and module washing is available

Construction manpower and unskilled and skilled labor availability

Construction contractors are available with machinery

Land is not a water logging area and natural rain water drain slope is available

Power evacuation route and GSS is nearby and approachable

Arrange land survey, contour survey, soil investigation and soil resistivity.

Keep special watch for Right of the Way (ROW) for all aspects.

Arrange Sun availability and weather data for last 20 years.

Find and mark the true North at site and on the drawing board.

1.2 Drawings

Based on the site visit report and the site plan of the land, designing work has to start on the drawing board. This report should contain the location co-ordinates, boundary line of the plant, contour survey, soil investigation reports and approach roads, shadow producing objects.

The best of surveyor should be employed to prepare this plan, as a one degree wrong plan, can result in the complete design going astray.

The first step is to place the solar modules on the drawing board within the boundary, as given by the survey report. This is the most difficult task, though looks simple, because the drawing so developed must be perfect. Following aspects needs to be addressed-;

-modules to be truly aligned to the desired direction

-covering maximum area of land

-no land wastage

-no shadow area placement of the modules

-proper row to row gaps

-usable pathways

-proper location of inverter room

-proper placement of cable routes

-good location of auxiliary power supply source

-use minimum quantity of cable to reduce cost and losses

-easy and shortest route for power evacuation

-proper use of natural slope of the land to reduce civil works cost

And many more aspects-therefore the designer has to consider the sizing of the plant in a very professional way, to satisfy the demands of the investors, site construction engineers and comparisons with other plants in the area. Therefore he has to follow the path, as detailed in the write up here below, to finally be in a position to come to the drawing board to design the proposed plant.

1.3 Check system loss Details at the proposed site. The PRACTICAL loss chart, being considered world over, is as below-;

System Loss Details		
Thermal Loss Factor (L1)	%	-9.9
Loss due to irradiance level (L2)	%	-1.6
Soiling Loss (L3)	%	-1.5
DC Wiring Ohmic Loss (L4)	%	-1.5
Module Quality Loss (L5)	%	-1
Module Mismatch Losses (L6)	%	-1
Inverter Efficiency Loss (L7)	%	-2
AC Ohmic Loss(Inverter To Transformer) (L8)	%	-0.2
Step-up Transformer Losses (L9)	%	-1
HV Transformer Losses (L10)	%	-0.5
Transmission Line Losses (L11)	%	0
Total Losses (L1- L11)	%	-20.2
Performance Ratio	%	79.8

1.4 Practically the losses occur in sections-; Starting from –modules- to cables- to SCB–to cables- to inverter- to transformer- to cables- to transmission line and finally to GSS. We make a calculated capital cost V/S losses analysis to arrive at the best combination, so that project cost is not un-necessarily increased to save minor losses. A balance is to be made between the two, to get best generation output.

1.5 As per our design, we have practically achieved the losses as below-;

1. Module to SCB-ohmic loss ....0.18% Calculated Max.
2. SCB to Inverter-ohmic loss ....0.23 % Calculated Max.
3. Inverter at 50deg. Cent. Conversion loss ....1.4 % Measurable
4. Transformer at 75% loading ....1.0% Measurable
5. Transformer output to GSS......................... 0.8 %. Measurable

TOTAL...3.6 % for 1 MW.

1.6 Performance Ratio (PR) and Capacity Utilization Factor (CUF) of a PV plant-;

Both the factors Performance Ratio and Capacity Utilization Factor are important for measuring the performance of a pv plant.

There is a debate among the designers as to which factor is more correct, and should be referred to while designing a PV solar plant. My view is, in Indian plains, we cannot neglect any of the above two parameters for designing and for assessing the performance of a PV plant. PR, CUF, STC, and NOCT have significant effect in PV plants and each factor, has to be given its due importance.

1.6.1 Performance Ratio- Performance ratio (PR) is calculated as a percentage and describes the relationship between the real and the theoretical/possible energy output of a solar PV plant.

Performance ratio gives insights of the complete, actually installed plant as to how efficiently; the available solar energy is converted into electrical energy. Deviations in the PR (e.g. values below the expected range) indicate a possible fault or problem in the solar PV plant. Decline in PR should be analyzed daily and reasons for this decfline should be recorded in the daily log sheet. This dcline in PR is therefore a warning about the plant deterioration or problems in the equipment's leading to outages or low efficiency and also inadequate Operation and Maintenance. The closer, the PR value for a PV plant approaches 100 %, the more efficient the respective PV plant is operating.

PR can be calculated on Daily/Weekly/Monthly/Yearly basis to continuously monitor the plant performance. As per the, so far achieved study, an average PR of 75% per year is a very good achievement.

CUF is the ratio of the total energy actually generated in kwh, compared with the theoretical energy in kwh that could have been generated in 365 days (8760 hours). It does not consider the no-sun time of the day of 24 hours.

The CUF does not take into account any environmental and loss factor, like variation on irradiance, the de-rating or degradation of the panels, losses in the complete system, losses due to temperature rise, ohmic losses, grid outages, equipment outages, etc.

But in both cases we have the actual energy generated in the numerator of the formula, and this actual energy generated value is based on all considerations-design, irradiance, losses, temperatures, any other factor that influences the generation. Therefore both factors have a significant status in the study of the performance of the plant.

1.6.2 Method of calculation of PR-;

The following formula is used for calculating Performance Ratio-:

PR=Actual Energy Generated/Irradiance*Active area of all PV modules*Efficiency of modules.

Following values are needed to calculate the performance ratio of a PV plant.

i).Firstly, Actual energy generated/exported to grid, measured by energy meters in **KWH.**

ii).Secondly, the irradiation data, measured by an irradiance sensor (Pyranometer) in **KWH/meter square**.

iii). Thirdly, the total active surface of the panels which receives the sun energy (calculate by multiplying the area of one module with total number of modules installed in the plant, in **meter square)**

iv).Fourthly, the guaranteed efficiency of the solar modules/panels given by the module manufacturer which is normally given as **15%.**

v). It is a must that the PV modules and the irradiance sensor are exposed to the same quantities of solar irradiation.

vi) The worst part of PR is, it does not indicate whether the basic design is under calculated or over calculated to decide the total no of solar modules per MW. It only analyses the existing/constructed/actual plant without indicating whether the no of modules installed and the capacity of inverters and transformers are correctly estimated and are for what capacity of the plant.

PR automatically accounts for all losses in the complete power generation system by taking actual value of units generated which are influenced by many factors, main being the temperature and low sun light.

Finally-**If the PR is 75%- it means that 25%of the solar energy from the sun, received by installed solar modules, was not converted into electrical energy.**

The following factors automatically influence the PR value in terms of reduced generation-:

- ✓ Temperature of all the equipments in the plant--------Design to keep them cool for higher PR
- ✓ Solar irradiation---you do not have any control.
- ✓ Pyranometer is shaded or soiled-------------------------keep it fit and clean
- ✓ PV panels are soiled---do regular water washing and
- ✓ PV panels are shaded--------------------------------------design for longer sun available window
- ✓ Ohmic losses--------------------------design for low I square R losses in AC and DC cabling.
- ✓ Efficiency of the PV modules/Inverters/Transformers---procure the best
- ✓ Plant Outage, partial or total ---------------------------- reduce down time
- ✓ Availability of the grid-----------------------------------ensure good grid availability

1.6.3 Method of calculation of CUF-;

The following formula is used to calculate the CUF .

CUF=Actual Total Units Generated (KWH) / Installed Plant Capacity (KW)*8760(Hours)

So, on one side, PR is a measure for the performance of a PV system, taking into account environmental factors (temperature, irradiation, losses etc.) and on the other side is CUF that completely ignores all these factors because it has installed plant capacity in its denominator.

CUF, in another form, is calculated for other non renewable generators, like thermal power projects, hydel, DG set etc. There this term CUF is called "Plant Load Factor" (PLF).

But in spite of above deficiency in CUF, we cannot neglect the importance of this factor. The CUF directly hits on the basics of the PV design and points out whether the designer has correctly taken the installed capacity of the plant in actual aspects.

There is a question and food for thought for the designers-"In solar PV plants, we are getting an annual PR value as 76% but the CUF is coming as 14%." Why?

Here the terms "PRACTICAL CONDITION CAPACITY" which is "NORMAL OPERATING CELL TEMPERATURE" (NOCT)of-;

- ✓ MODULES,
- ✓ INVERTERS, and
- ✓ TRANSFORMERS

....come into picture. These terms are explained in the coming chapters of this guide.

Can the PR go above 100%? Answer is YES-"laughable"

You cannot convert even the total energy of sun into electricity-what to talk of more than 100%. The explanation is-; the performance ratio is a variable which, under the influence of certain factors, may even exceed values of 100 %.

The partial or complete placing in shadow of the measuring gauge can result in PR values of over 100 %. In addition, environmental factors such as snow, dust, fog, smog or pollen can lead to soiling of your PV plant and thus also result in PR values of over 100 %.
Further-for all calculations of PR we are taking the module conversion efficiency as 15% in the denominator. An increase of 1% in module efficiency, making it 16%, reduces the PR by 5% and a decrease of 1% in module efficiency, making it 14%, increases the PR by 3%

1.6.4 PV Syst and Meteonorm study to check the generation level at the location –
This is absolutely necessary to prepare a PV Syst report for the plant at the selected location and this is the first step in designing a PV solar power plant. We need to know the longitude and latitude of the location so that sun position chart is prepared and a study is conducted to calculate the month wise generation estimate at the location.
PV Syst and Meteonorm software is an internationally accepted tool/software to plan the generation level at any location all over the globe.
Knowledge of this software and its operation to get the required data is the first step in designing of a PV Solar power project. Once a PV Syst and Meteonorm report for any location is made, we get from these reports the following-;
1. Expected Performance Ratio at the location
2. Month wise estimated unit generation at the location
3. Based on the proposed plant capacity in KW or MW, energy generation in each month is assessed.
4. Loss factor at the location based on the effective irradiance on the solar collectors i.e modules.
Based on above explanations the placed below is the PV Syst and Meteonorm report for a location in India.
A study of the Meteonorm analysis will indicate that the solar irradiance at the tilt angle of the report is on higher side. The reason being the data is based on the past 20 years of the study of solar data. We have practically seen that the value of irradiance as given does not appear actually nowadays. It is lesser practically. Therefore the field is open to the designer to increase the DC capacity and provide more number of modules to counter the low generation because of low irradiance. This will increase the cost marginally one time, but this additional cost is recoverable from enhanced generation during next 25 years.
Placed below is an exercise to find generation levels and associated data using above detailed tool.

Grid-Connected System: Main results

Project : **Grid Connected Project in Banda,U.P**

Simulation variant : **New simulation variant**

Main system parameters	System type	**Grid-Connected**		
PV Field Orientation	Seasonal tilt: summer/winter	5° / 40°	azimuth	0°
PV modules	Model	CS6X - 300M	Pnom	300 Wp
PV Array	Nb. of modules	3340	Pnom total	**1002 kWp**
Inverter	Model	Sunny Central 900CP XT	Pnom	900 kW ac
User's needs	Unlimited load (grid)			

Main simulation results				
System Production	**Produced Energy**	**1473 MWh/year**	Specific prod.	1470 kWh/kWp/year
	Performance Ratio PR	78.1 %		

Normalized productions (per installed kWp): Nominal power 1002 kWp

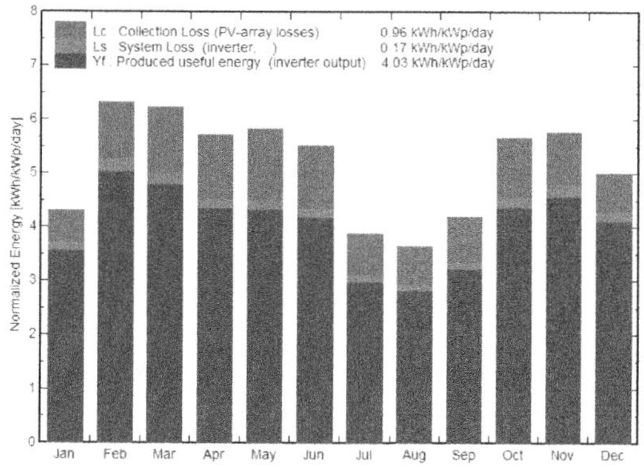

Performance Ratio PR

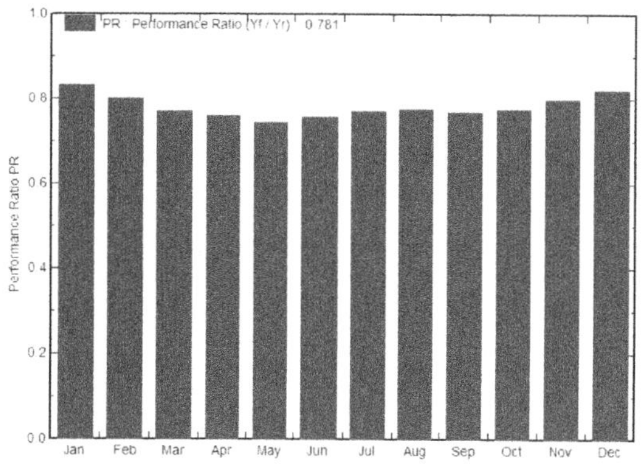

New simulation variant

Balances and main results

	GlobHor kWh/m²	T Amb °C	GlobInc kWh/m²	GlobEff kWh/m²	EArray MWh	E_Grid MWh	EffArrR %	EffSysR %
January	102.0	15.00	133.1	129.9	116.0	111.1	13.60	13.03
February	139.0	19.10	176.6	172.5	148.3	141.7	13.10	12.52
March	177.0	24.80	193.1	187.6	156.0	149.3	12.60	12.06
April	170.0	30.60	171.6	166.0	136.1	130.6	12.37	11.88
May	181.0	32.90	180.3	174.5	140.1	134.6	12.12	11.65
June	167.0	32.30	165.6	160.1	130.7	125.7	12.31	11.84
July	121.0	30.00	120.6	116.0	96.8	93.1	12.54	12.05
August	113.0	29.70	113.1	108.6	91.3	87.8	12.60	12.11
September	124.0	29.00	125.9	121.4	100.9	96.9	12.51	12.01
October	149.0	26.40	175.2	170.9	142.0	136.0	12.64	12.11
November	126.0	21.30	172.9	168.8	144.3	138.1	13.03	12.47
December	107.0	16.60	155.0	151.4	133.5	127.7	13.43	12.85
Year	1676.0	25.67	1883.1	1827.6	1536.0	1472.7	12.73	12.20

Legends:				
	GlobHor	Horizontal global irradiation	EArray	Effective energy at the output of the array
	T Amb	Ambient Temperature	E_Grid	Energy injected into grid
	GlobInc	Global incident in coll. plane	EffArrR	Effic. Eout array / rough area
	GlobEff	Effective Global, corr. for IAM and shadings	EffSysR	Effic. Eout system / rough area

PVSYST V5.74		28/11/14	Page 3/3

Grid-Connected System: Loss diagram

Project : **Grid Connected Project in Banda,U.P**

Simulation variant : **New simulation variant**

Main system parameters	System type	**Grid-Connected**		
PV Field Orientation	Seasonal tilt: summer/winter	5° / 40°	azimuth	0°
PV modules	Model	CS6X - 300M	Pnom	300 Wp
PV Array	Nb. of modules	3340	Pnom total	**1002 kWp**
Inverter	Model	Sunny Central 900CP XT	Pnom	900 kW ac
User's needs	Unlimited load (grid)			

Loss diagram over the whole year

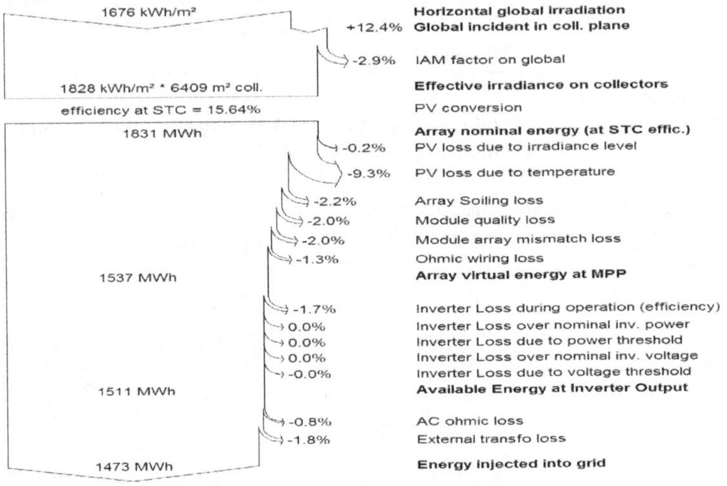

Activity 2: Regulatory issues based on solar policy of the government

In today's status the solar power generated can be sold to the DISCOMS and for this certain statutory norms are to be done as per the SOLAR POLICY OF THE STATE AND CENTRAL GOVERNMENTS. The

Above aspects are not covered in this guideline, because these are not the technical consideratio for designing a solar power park.

Activity3: Site engineering assessment

3.1 Civil Engineers Domain

The first step in preparing for the solar development on the land, involves the preparation of the site for construction. Site preparation includes clearing the site, filling and grading, and compaction.

First, all trees, bush, and overgrowth will need to be cleared. Precautions should be taken such that clearing equipment, such as bulldozers, does not scrape the soil cap. Using virgin land/soil for solar module mounting structure installation is the best thing-of course the land should be plain. The soft soil land should not be dug un-necessarily.

If a weed treatment is done at this stage itself, grass and plants will not grow in the area thereby saving the cost involved in weed cleaning every three months throughout the life of the plant.

The exact depth of the soil cap required will be determined during soil investing and the design phase of the project.

It is not necessary for the site to be graded to a zero slope, as a rising slope from south to north is acceptable, and would possibly allow the distance between rows of solar arrays to be smaller, due to the reduction in shading distance requirements. But in this case the pile height above ground, of each pile, has to be checked with water pipe level measurement. It is to be established that each pile top is in the same level as per ground slope.

String formation should be done in such a way that all strings come on the same level of land. This will remove all chances of higher loss of generation if shadow of one table falls on the other.

Use natural land slope to design the rain water drainage system. The study can be conducted during rainy season and the water flow natural routes are marked. These can be developed to let the water evacuation possible. The man made drainage system will be very- very costly and will not work in such vast land with random installations.

3.2 Photovoltaic Equipment Selection and System Design Criteria

3.2.1 Module Mounting Structures and Tilting of modules to face the sun perpendicularly

A tilt is the angle that the module makes w.r.t earth or horizontal plane, so as to face the sun perpendicularly or near to perpendicular. Tilting is aimed to make the modules to face/absorb maximum sun light directly throughout the day/season when the sun position in the sky changes from sunrise to sun set and winter to summer and back to winter.

In general, there are four types of PV structure support system configurations being utilized in large scale solar plants world over, which are **fixed tilt, single axis tilt, dual axis tilt and seasonal tilt.**

Axis tilt type arrangements need continuous movement of module to face the sun from sunrise to sun set and from winter to summer.

The fixed tilt is self explanatory.

The seasonal tilt has to be done manually in every season such as summer, winter and the periods in between them, both ways.

The comparison between fixed tilt and seasonal tilt systems indicates-;

- Fixed tilt systems are lighter in weight, are less expensive, and produce less energy per installed kW on an annual basis than a single axis tracking system;
- Single axis and Dual axis tracking systems are heavier (requiring deeper piles for footings), more expensive, and produce more energy per installed kW on an annual basis than a fixed tilt system. But they require auxiliary power for the synchronous motors, which makes it more costly. The auto electrical system, to arrange tilt, is not a success in Indian climate because of heat and dust.
- The claim, that the axis tilt system increases the generation by 30% is not confirmed and proven so far in Indian climate.
- Nowadays people are going in for manual seasonal tilt, by adjustment of angle of module mounting structure during summer and winter. It is proven now that this increases the generation by 15 to20% extra.

3.2.2 Methods of foundations of MMS-The cost of civil foundation work are the main issue, apart from the strength of foundation, to withstand all the forces that the structure will face when fully loaded. We have pile foundations where we drill a hole in the ground, place the leg inside it at the required depth and fill the side all around it with cement mortar. In soft soil, this arrangement is not a success. So there, we will have to go in for heavy foundation, which are costly. Type of foundation depends on the type of soil.

3.2.3 The MMS can be designed for various configurations, such as single portrait, double or triple portrait or single/double/triple landscape or any other combination. The main considerations are the strength of the structure to face the wind forces and load of the modules, the quantity/weight/cost of the total steel structure, and savings in the land area.

In view of above explanations the task of designers becomes very complicated as they have to look towards all the above aspects for a very efficient and economical PV solar power plant. Placed below is a sample of the module mounting structure, which is indicative, to give an idea of what the MMS is like. There are a number of such combinations/designs all over the world, and engineers keep on modifying them to make them more cost effective, simple yet strong and easy to manufacture, transport, handle, install, and maintain.

Present day achievement is 41Tons/MW as the steel quantity used in MMS.

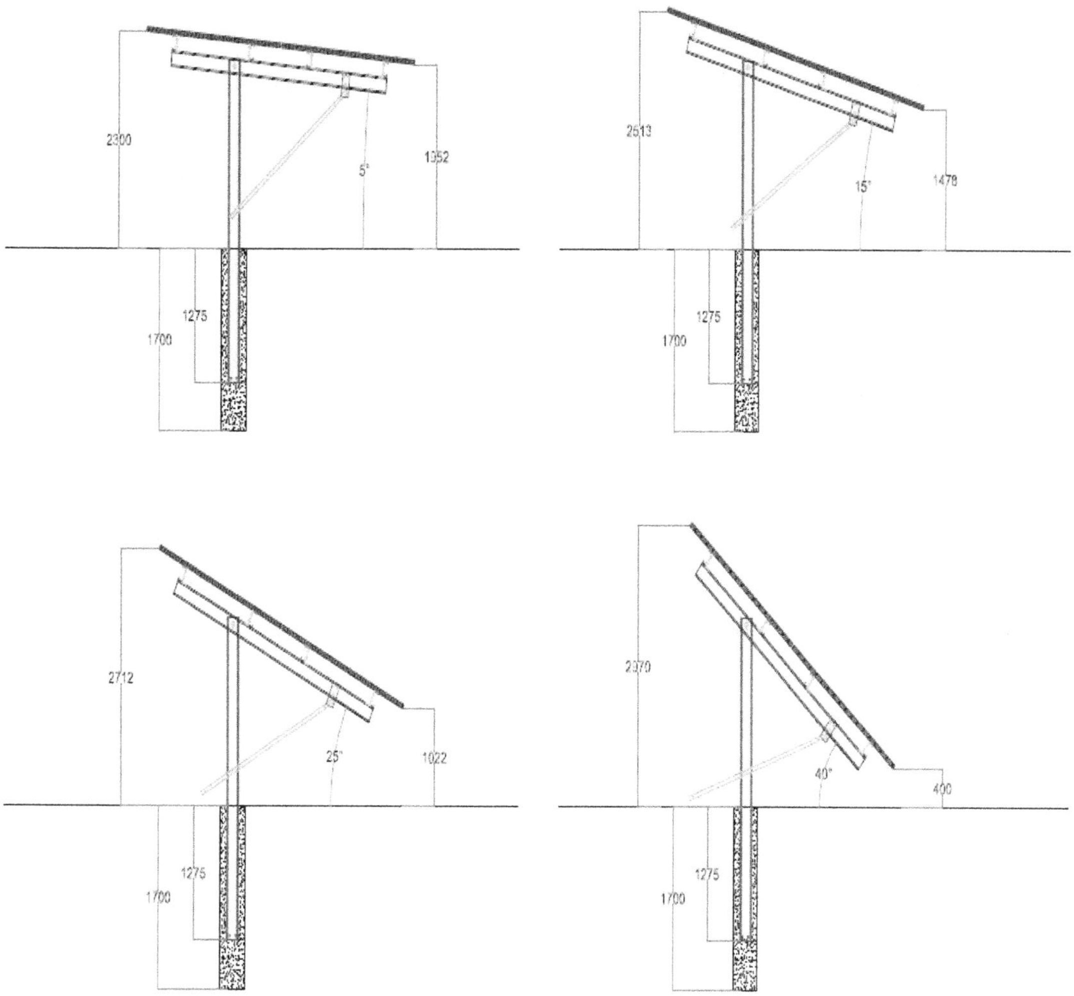
2300
5°
1952
1275
1700
2513
15°
1478
1275
1700
2712
25°
1022
1275
1700
2970
40°
400
1275
1700

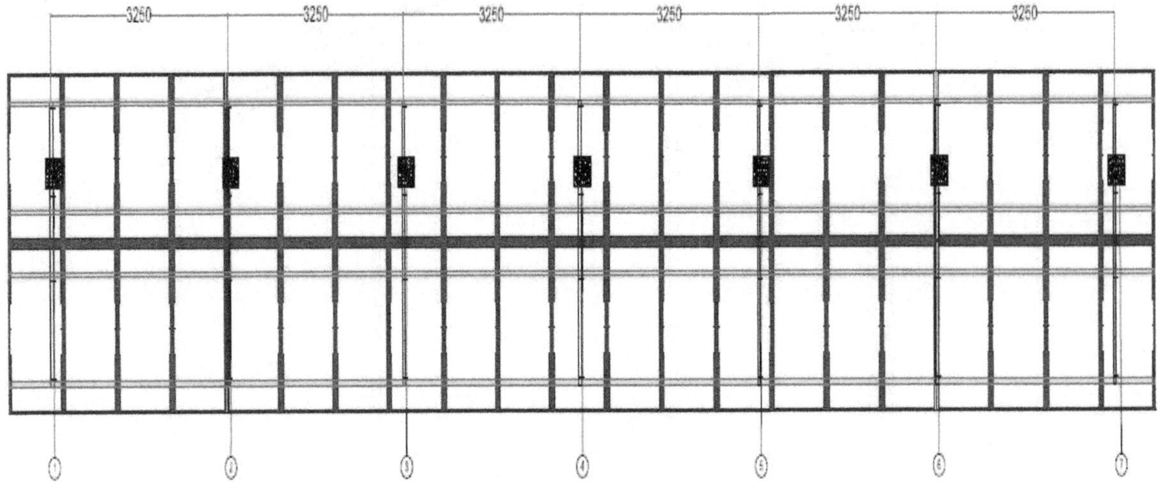

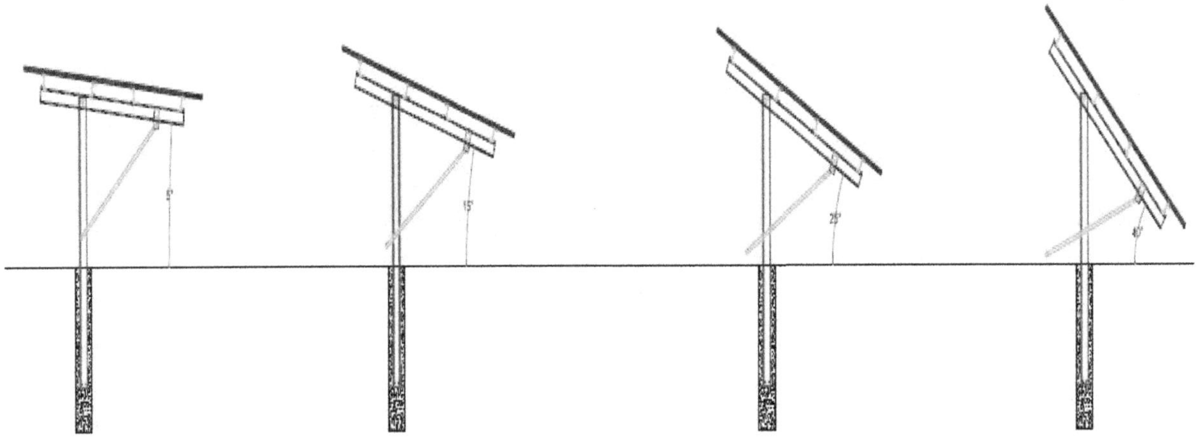

The figures above show the complete structure details of a seasonal tilt support structure along-with their foundation details-the fig. is self-explanatory.

3.2.4 Module Mounting Structures-constructional details-;

We can divide the total module mounting structure in two parts-;

Part 1. It is the fixed support embedded in the ground-it is called pile.

Part 2. It is the superstructure fixed on the pile and it holds the modules.

The module holding structure is provided with manually/electrically operated arrangement of tilting the modules with change of season from summer to winter and from sunrise to sun set.

The most common design for the systems is a pile configuration wherein holes are drilled in the soil, and concrete footings poured in the hole in which pile is held.

The depth of the pile is determined by the dead weight point loads supported by the pile and footings.

While doing piling work, always check the level of each pile with water tube level monitor.

The pile strength and depth for a mounting system is based on the soil investigating report, which gives the permissible loading limits of the land in question. Civil engineers actually install the pile and then do the pull out test to see that the foundation design done is strong enough to withstand all forces on the erected structure.

3.3 Forces on structure assembly

Apart from the weight of the module mounting structure itself, the weight of the modules is the basic criteria for the strength of supporting system in totality. The vertical load centre of gravity factor is to be checked by the structure designer.

The final design criterion to be considered is wind loading. The Indian adopted practice is to design a system to withstand 120 mph winds. The probability factor is decided by checking the past 20 year's history of storms in the area.

This is primarily addressed through specification of mounting system structure components, namely piles, racks, bracings, cleat to hold the module and perlins of significant gauge steel to withstand all loadings and forces.

Mounting the panels close to the ground reduces the system weight and as a result reduced pile depth requirements apart from saving quantum of damage during storms. Based on this general engineering assessment, it is recommended that the site be developed with fixed tilt mounting base pile with provisions of employing tilting arrangement in the module supporting structures.

3.4 PV MODULES -;

There are two general types of PV panels with numerous variations within each type with respect to size, shape, substance, and cell materials. These are mono/poly-silicon/ crystalline and thin film.

Mono- or poly-silicon/crystalline are higher efficiency (produce more power per unit area) panels, but typically weigh more and cost more on a dollars per watt basis.

There are many considerations for and against, for both the types as summarized below.

Since modules made from crystalline silicon are generally more efficient than thin-film modules, they are used wherever space is at a premium.

In the case of crystalline solar modules, an area of around 7 bigha (165X165 feet) is needed to achieve an output of one MW. Whereas, for thin-film modules the area required for the same output is 11 bigha,

This means that the cost of support structures and installation is higher for thin-film solar modules, as surface area efficiency is lower, and that the modules themselves must therefore be cheaper in cost comparison.

Crystalline silicon solar cells are particularly responsive to long-wave solar radiation.

In contrast, thin-film modules make better use of the short and medium-wave range of the solar spectrum.

In cloudy conditions, the spectrum, that hits the ground, has a higher proportion of shortwave light, which is best exploited by amorphous thin-film modules.

In general, thin-film modules are ideal for sites which experience a high proportion of diffused insolation due to frequent cloudy weather, severe pollution, dirt settlement on modules frequently, or temporary or partial shading. It is for the designer to take a call about selection of type of module depending on the site conditions and cost aspects

3.5 The temperature effect on PV modules-;

Insolation can heat PV modules to as much as 70 °C. For this reason, they are installed so as to ensure that air can circulate to provide sufficient rear ventilation. Only natural cooling with wind is possible, so place the modules in such a way that natural air flow is not restricted.

As the temperature increases, the PV module output drops steadily. The module suppliers specify the temperature capability of modules and the related loss of output/efficiency at different ambient temperatures in open sun area to which the modules are actually subjected.

This makes the task of the designer very difficult, he faces the dilemma as to on what ambient temperature he has to design the total number of modules for the required kw/mw power capacity. More so, because the temperatures, to which the modules are actually subjected during operation, vary drastically, from summer to winter, (0 deg. centi. to 50 deg centi.) and also from sunrise to sunset. He has to take a medium path and design the system in such a manner that the power generation at STC conditions does not overload the connected DC power system consisting of cables and modules. We have discussed this aspect in coming chapters ahead.

Crystalline modules (cSi) are far more severely affected by this, than thin-film modules (aSi and CdTe). At a module temperature of T=55 °C, the solar module would show a drop in output of 9%.

3.6 Bypass diodes in the module-Whenever a cell in the module gets covered by a shadow or is damaged by any other reason, it stops generating, but since it is connected in series with other cells, it will give resistance to the total series current and will reduce the output from the entire series. The reduced output and possibility of damage to cell and/or module itself can be reduced by the use of bypass diodes. The by-pass diode short circuits the affected cells inside the modules and allows the current to by-pass the damaged cell.

Since a single solar cell is only able to generate around 0.5 volts a number of cells are connected in series to form a module. This makes the module extremely sensitive to partial shading, because when a shadow is cast on a cell, the cell can no longer generate power, turning it from power generator to power consumer. As the weakest link in the chain, connected in series, it restricts the power output of the entire module.

Shaded cells or internally damaged cell do not generate electricity, while the other, fully illuminated cells in series remain completely active and drive their power through the shaded cell, which in turn offers resistance to the current thereby converting that power into heat. Finally it leads to a "hot spot", which will ultimately melt a hole in the cell material. Same is also true for a cell damaged due to any other reason

A bypass diode, which by passes the cell string containing the shaded/damaged cell, is therefore used to steer the electricity past the passive cell. A bypass diode usually bypasses 20 to 24 cells. There are about 70 (and above) cells in a module.

As each diode by passes a part of the module, in the case of shading/damage, only some of the output of all the series-connected cells making up the module will be lost. It would therefore be ideal if each solar cell could be equipped with a bypass diode. But this will make the cost of the module very high. Therefore module manufacturers opt for a number of by-pass diodes for a group of cells inside the modules.

3.7 Blocking Diode to block reverse current flow-With integrated grid tied operation of the modules, there are chances that there may be a reverse power flow into the modules. This reverse flow of energy will damage the module i.e. the current generator.

We can compare it with the motoring of the grid connected generators, if the prime mover stops to feed mechanical energy, then the grid power runs the generator as a motor.
Actually the Blocking of reverse current facility should be in built in the module itself. But this will make the size of the module very big and the total internal circuit of the module will be very cumbersome and more prone to faults. Therefore this is not provided in the modern day modules. So we will have to provide this facility in the SCB-but here again the SCB will get heated as the Diodes get heated during loading.
Therefore we provide fuses in both the –ve and +ve of the strings (made of 20/22 modules) that are connected to the bus of SCB. These fuses should be rated equal to the Isc of the module. During any chance of flow of reverse current-this, in all cases, will be much more than Isc of module. This hi current will blow the fuses so provided. The fuse protects bi-directionally, and will work as Blocking Diode. These fuses are located in the String Combiner Boxes.
The O&M engineers should study the above two aspects during running of the plant. This phenomenon is now appearing in the plants and damaging the modules by internally damaging the cells.
Causes for failure of modules- The above two arrangements protect the damage to modules during operation, still the modules fail. Fundamental cause of module failure is manufacturing defect in the tiny cell. Shadow on single cell can also happen due to bird droppings or cement spread during construction.
Since we have provided protection in the form of By-Pass Diode and Blocking Diodes, and the fuses in the SCB, there are no chances of reverse power flow in the modules or strings, even if there is shorting of SCB outgoing +ve and –ve cables. This will blow the fuse.

3.8 Module internals-A module is a series connection of small cells, each cell developing 0.5 volt and 2 watt power-presently these values have increased to 1-volt and 4-watt.

- Module junction boxes connect solar cells to the outside world by joining the connection cables of the cell strings and interconnecting them with the bypass diodes and the module connection cables.
- To prevent moisture from entering the module junction box, it is waterproofed and often sealed with silicon.

  Practically the solar module design has reached its final shape, and there are very little chances of any future development. Of course – increasing the individual cell voltage, reducing the size and weight of solar modules are under experimentation. The life expectancy of modules (25 years) is under challenge and manufacturers are experimenting on this aspect.

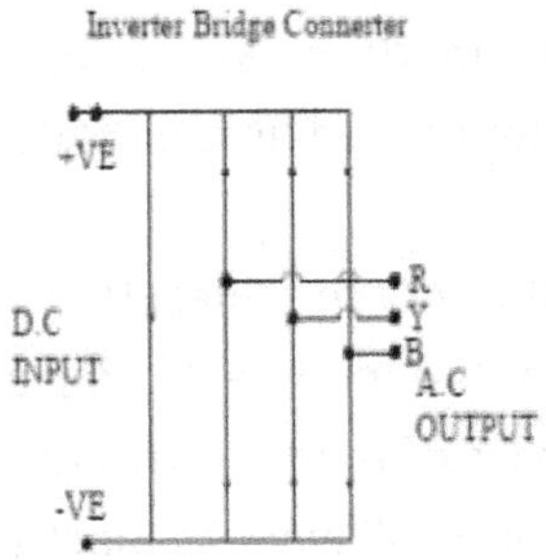

Fig-1

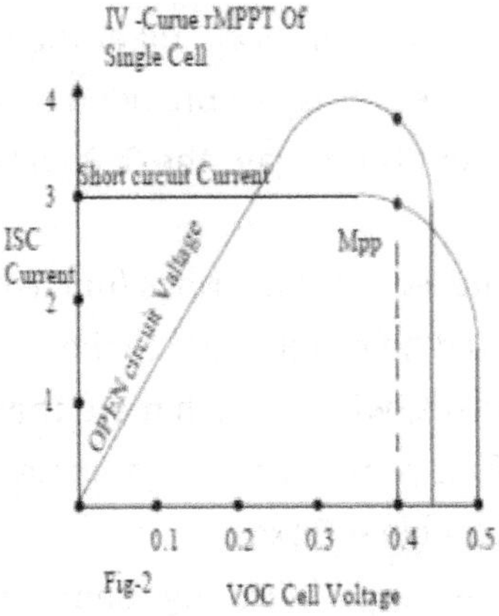

Fig-2

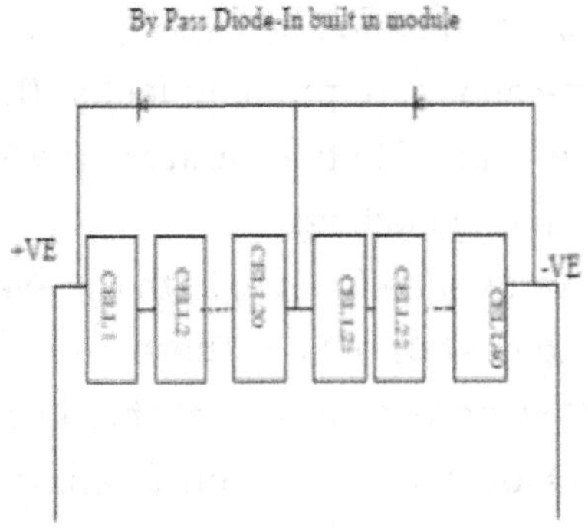

Fig-3

Fig. DC to AC converter Bridge-Cell Voltage-In built bypass diode

Activity 4: Plan PV system size, conceptual design elements, and specifications

4.1 PV System Size Options - A 1.0 MW solar farm is the desired size for this study. A solar farm of this size utilizing amorphous silicon modules will require approximately 3 to 5 acres/7 Bigha of land at the site. Whereas, for thin film the land required will be 7 acre/ 11 bigha.

4.2 Conceptual Design Elements

4.2.1 The type of PV panels and mounting structures define the major conceptual design elements of the solar farm.

Maximizing the efficiency of solar installations hinges on the ability to install as many panels as possible in a given area. Often, the question of whether to install the panels in landscape or portrait orientation arises. More modules can be installed in Portrait system for a given land.

Common thought suggests that, reason for choosing one orientation over the other, is the number of panels that can be installed in a given length or the ability to install more panels over a given footprint.

The mounting racks would be aligned in rows along an east-west axis across the entire area defined for the project.

Depending on the height of the panels off the ground, it is estimated that approximately 3 to 4 meters of spacing between rows would be required to prevent shading from one row of modules onto the other. Much depends on whether you use the portrait (single/double/triple) or you go for landscape method of placement of modules.

This row distance is determined by the "design day," or "Sun available Window" on 22nd Dec. from 9.00AM to 4.30PM, when the sun is at its lowest point, above the horizon and casts the longest shadow in India.

In addition, providing adequate distance between rows allows for access by maintenance vehicles or small cranes, should system maintenance or module replacement is required. Therefore you do not need to provide extra pathways for the movement of O&M staff and small vehicles.

4.2.2 Main Equipments of the Solar Power Plant–

The main equipments of a MW capacity solar power plant are-:

- solar modules,
- module mounting structures,
- cables,
- String combiner box,
- inverters,
- transformers, LT and HT switchgear
- four pole structures,
- meters and the power evacuation line.

In addition, the solar farm developer will need to work closely with DISCOM AND TRANSCO to meet all interconnection requirements for connecting with the power grid.

The following is the guidelines and criteria for selection of the above detailed main equipments of the PV solar power plant.

4.2.3 Inverters, also called as POWER CONDITIONING UNITS as they receive DC power and convert it to AC power supply -; the job of the inverter is to convert the DC input power to AC output power and connect this system with the DISCOM grid to inject solar power into it with the highest possible efficiency. A PV installation's yield is, therefore, dependent on the reliability and efficiency of the inverter.

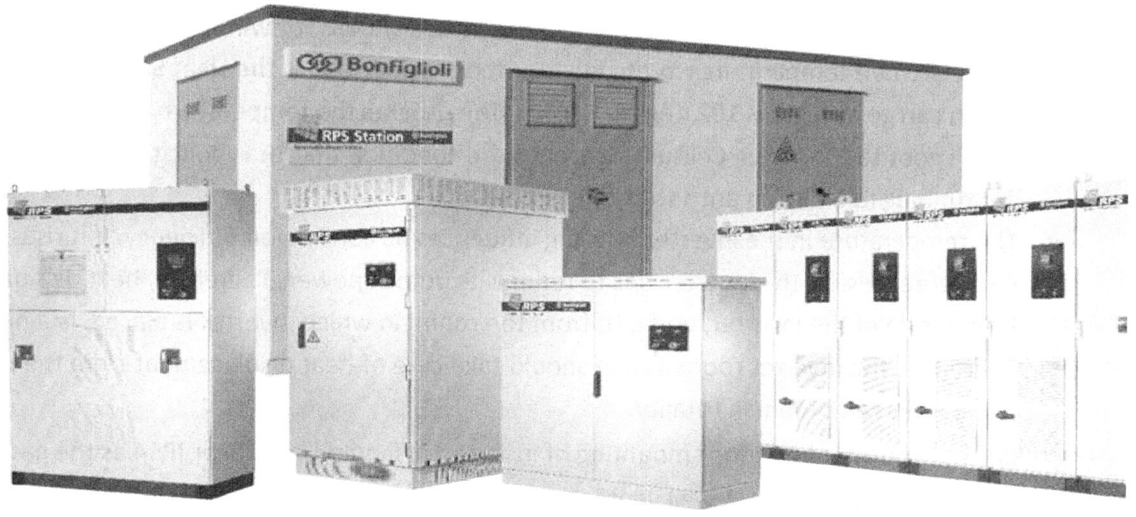

Photograph of various sizes of inverters.

To calculate capacity of Inverters for grid tie systems or grid connected systems, the input rating of the inverter should be same as PV array rating, to allow for safe and efficient operation. The reason is, the modules have a tolerance of +3% to +5% to take care of sun flares, low ambient temperatures and good upkeep. But no overloading of the inverters is technically permissible. If, by any reason more DC power is put-in into the inverter, it will clip the DC input current and raise the DC input voltage because of the electrical fundamental law-Power=Voltage X Current

Inverter also controls and monitors the power generation of the entire PV system connected to it via the SCB.

The tasks of a PV inverter are -;

- ✓ The most important characteristic of an inverter is its conversion efficiency. This value indicates what proportion of the energy "inserted" as direct current, comes back out in the form of alternating current, operating with an efficiency of around 98 percent. Of course subject to the temperatures. The limit is 45°C, but we insist on no loss operation of inverter up to 55.
- ✓ The power characteristics curve of a PV module is strongly dependent on the radiation intensity and the temperature of the module. In other words, on values that continually changes over the course of the day.
- ✓ But when we connect large number of strings in SCB and many SCB in one inverter, the generation level hunting is consolidated at the inverter end itself.
- ✓ For this reason, the inverter must find and continually observe the optimal operating point on the power characteristics curve, in order to "bring out" maximum power from the consolidated system, in every situation.
- ✓ The optimal operating point is called the "maximum power point" (MPP), and the search for, and tracking of, this MPP is correspondingly called "MPP tracking."
  The inverter monitors the energy yield of the PV plant and signals any problems.
- ✓ It also monitors the power grid that it is connected to. Thus, in the event of a problem on either side, it must immediately disconnect the plant from both ends, as the case may be.
- ✓ Islanded mode of operation is a must for the inverter. More so, because PV modules are always live when light is shining on them, they cannot be switched off.
- ✓ Communication interfaces on the inverter allow control and monitoring of all parameters, operational data, and yields. Data can be retrieved and parameters can be set for the inverter via a network connection, industrial field bus such as RS485, or wireless.
- ✓ There are two temperatures in the inverter monitoring- One is the Heat Sink temperature, which can go as high as 102 degree Celsius. The other is the temperature of inverter internals, which goes to 55 degree Celsius, subject to the condition that the cooling fans of the inverters are running full load on auto start mode.
- ✓ The temperature in the inverter housing influences its conversion efficiency. If it rises beyond 45 degree Celsius, the inverter has to reduce its output power. Therefore, heat dissipation from inside of the inverter, and also from the room, in which inverter is placed, is a necessity. Therefore, the inverter room design should take care of heat displacement from the inverter in particular and room in totality.

Ingress Protection-; for indoor mounting of inverter, it should be IP20, or IP54 as the case may be and for outdoor it will be IP65. A well ventilated, weather-proof enclosure, ideally built in line with protective rating IP, is required.

4.2.4 Power Transformer Step up transformers are used to connect MW scale PV plants to the utility network, their sizing being often accomplished by taking into account the PV plant peak power. Transformer ratings should be such that it is neither very high nor very low. A high value will lead to increased cost and losses and a low value will restrict the full power flow.

There is a simple design short cut/thumb rule for arriving at the capacity of the inverters and transformer-; The DC capacity of modules in KW, should be equal to AC capacity in KVA of the inverter, following same logic, the kva rating of transformer will be same as kva rating in AC of inverter. For one MW DC capacity of a plant the inverter rating on AC side will be one MVA and transformer rating too will be one MVA.

But in doing so we should not be fooled by the module and inverter suppliers who indicate higher capacity at "Standard Test Conditions" and less at 50 deg. They sell their modules and inverters stating that the capacity of inverter is as per STC which is higher than NOCT. This aspect is explained in detail in the coming chapters.

Transformers are static devices, and can be overloaded continuously, up to about 10% of its capacity, of course subject to the limitations of the temperature rise of oil up to 50 deg. cent. above the ambient and short term overloading permissions are still higher than 10%. Designers take advantage of this overloading permission and reduce the ratings of transformers to save cost. Issue is open to discussion and has many ifs and buts.

We select the commonly used oil filled power transformers with OFF CIRCUIT TAP CHANGER. The voltage transformation ratio depends on the output voltage of the inverter, and the grid voltage of the DISCOM. For this tailor made transformers are to be got manufactured after settling the system parameters. Conventional protections-Bucholtz, OTI and WTI are required to be there.

4.2.5 HT Circuit Breaker and Protections

Figure shows the actual picture of the HT panels placed in the control room

In any electrical circuit, whether it is a high voltage power evacuation feeder, or a low voltage power carrying circuit, or a domestic low voltage supply system, the importance of the circuit breaker cannot be overlooked. The main function of the circuit breaker is to protect the plant, as well as the persons/plant operators working on the system. In case of a ground fault or short circuit fault in the system, the current and/or voltages of the faulty system rises to infinite levels, causing heavy damages to the system, and operators may also get electrocuted.

As the electrical fault should be cleared within 3 to 4 milliseconds, it is not possible for the operators to disconnect the faulty section manually. This is very risky also. So the necessity of the circuit breaker will never be neglected by any designer. But there is a possibility that one breaker is provided to cover four or five plants. With this arrangement we can have a common protection for a group of plants situated nearby. Present trend is to go for a group protection system. This reduces the cost of the plant. The VCB should be complete in all respects, with protections of transformers, earth fault, over current and over voltage etc.

In the TABLES all over this book, what is important is the TABLE itself. The values/data filled therein is for specific equipments and is indicative only. This data will vary from site to site or plant to plant.

S.NO	Tech-specs. Circuit Breaker for 33 kv	
1	Rated Voltage	36KV
2	Rated Current	1250Amps
3	Frequency(Hz)	50+-5%
4	Insulation Level	70(KV)/170(KV) Peak
5	Short circuiting breaking current	25KA rms
6	Short circuit withstand current and duration	25KA for 3 Sec
7	Short circuit making current	62.5KA
8	Operating sequence	O-.3S-CO-3Min-CO
9	Closing coil supply voltage	24V DC
10	Opening coil supply voltage	24V DC
11	Motor Supply Voltage	230V AC
12	Should be SCADA complaint	
13	IP(Ingress Protection)	54
14	33KV breaker should be equipped with the Multi functional, Microprocessor based, numerical relay.	
14	**Relay Protections**	
A	IDMT Instantaneous over current/short circuit	50/51N
B	IDMT Instantaneous earth fault protection	50/51N
C	Breaker failure protection	50BF

D	Trip circuit supervision	74TC
E	Lockout	86
F	Overload Protection	49
G	Standby EF protection if available	51G
H	Under and over voltage protection (optional)	27/59
I	Under and over frequency protection	81 O/U
J	Directional determination for Over- current, phases and earth fault sequence. Any other protection, which may be required during detailed engineering, is to be included	
15	**ALL Transformers will have following protections:-**	(transformer mounted via marshalling box)
A	Bucholtz trip	(63T) -
B	Oil Temperature trip	(26T) -
C	Winding Temperature trip	(49) –
D	Pressure Relief Device -	PRV

4.2.6 AC Power Circuit-Cables, four poles and O/H lines

The AC power circuit designing has to look after the following factors-;

Cables and conductors used should be rated to take the maximum possible load of current/ voltage/power that the system will feed in at any time.

In fact, to reduce losses and to look after de-rating of the conductors, we always select cables and conductors of the capacity much higher than the expected maximum power flow in the circuit.

The insulation quality of cables should be very good and should be safe in the type of atmospheric conditions. Always select the cables from known and reputed manufacturers.

The four pole design should be such that the maintenance, in case of faults is safe and easy for the lineman to work on. For clearances, please strictly follow the Indian Electricity Act 1953 and its revised version.

Four Pole Structure with CT/PT/Cable/LA/Power evacuation line etc.

Four Pole Structure with CT/PT/Cable/LA/Power evacuation line etc.

Activity-5-; General Technical Specifications of main equipments.

Based on above details, we can decide the technical specifications of the main equipments. However these are general parameters giving guidelines, but detailed specifications will be provided by the manufacturers and suppliers based on these guidelines.

For all equipments of the plant we should always insist on the following, very strictly-

Type Test,

Routine Test,

Acceptance Test,

Test Certificates,

Past Performance Certificates from end users and the internal

Quality Assurance Plan of the manufacturers.

5.1 Specifications of Solar Modules

Modules are to be selected based on the site conditions, solar radiations, type-thin film or Crystalline. Therefore the below data is to be collected from the manufacturer

a) Name of the manufacturer of PV Module

b). Name of the manufacturer of Solar cells

c). Unique Serial Number & Model No. of the module

d). Month and year of the manufacture (separately for Solar cells and module)

e). Country of origin (separately for solar cells and module)

f). I-V curve for the module
g). Peak Wattage of the module, Im, Vm and FF for the module
h). Date and year of obtaining IEC PV module qualification certificate
i). Name of the test lab issuing IEC certificate
j). Other relevant information on traceability of solar cells and
module as per ISO 9000 series
k).RFID shall be mandatorily placed inside the module laminate.
l) A distinctive serial number starting with NSM will be engraved on the frame of the module or screen printed on the teller sheet of the module.

5.233 kv Vacuum Circuit Breaker--Details are already given above.

5.3 Metering Arrangements

	Metering CT core type-Outdoor oil immersed for 33 kv	
S.NO	Particulars	
1	Higher System Voltage	36KV
2	Nominal System Voltage	33KV
3	CT Ratio	HV side As Required/5-5 Amps or 1/1 Amps
4	Class	.2s
5	Burden	15 va
6	Frequency	50Hz+-2%
7	Suitable for metering purpose	
8	Power Frequency withstand voltage	70KVrms
9	Lightning withstand voltage	170KV peak
10	IS	2705:1992
	Metering PT core type-Outdoor oil immersed	
S.NO	Particulars	
1	Higher System Voltage	36KV
2	Nominal System Voltage	33KV
3	PT Ratio	33KV/110-110V
4	Class	.2s
5	Burden	30VA
6	Frequency	50Hz+-2%
7	Suitable for metering purpose	
8	Power Frequency withstand voltage	70KVrms
9	Lightning withstand voltage	170KV peak
10	IS	3156:1992

S.NO	ABT – Import/Export Meter Technical Particular	
	Meter shall be suitable for (3Ph, 4 wire)	
1	Nominal System Voltage	33KV
2	Higher System Voltage	36KV
3	Auxiliary Supply Voltage	230Vav-+10% (220Vdc/110Vdc-/+10%)
4	Nominal Current	5Amp
5	Max Current	5Amp
6	Power Factor	0 to +-1
7	Frequency	50Hz+-2%
8	Accuracy	.2s
9	Humidity	95% non condensing
10	Environmental Specification	IP 54
A	Operating temperature	-5 to +60 degree Cent
B	Max ambient temp	55C degree Celsius
C	Max ambient temp in shade	50C Degree Celsius
D	Max wind pressure	150Kg/Mtrsqr
11	Meter shall be ISI mark/any PRACTICAL international mark	
12	IS	14697/1999
13	Meter shall be projection type suitable for mounting on plane vertical surface, dust and moisture proof	
14	IP(Ingress Protection)	54
15	Meter body shall be poly carbonate as per IS 11731	

5.4 TECHNICAL SPECIFICATION OF ----- kVA Transformer-SAMPLE

The transformer will be designed and manufactured as per IS 2026 and also CBIP guidelines, and will be supplied with first filling of oil to IS 335-1993.

Make: Name of manufacturer……………………………………

1.01	**Parameter**	**Particulars**
1	Rated KVA	1180X1
2	Winding Arrangement	2 winding
3	Type	Core Type-Oil Immersed
4	Type of cooling	ONAN
5	Winding	Copper Wound
6	Insulation Class	A
1.02	**SYSTEM PARTICULARS**	
1	Nominal Voltage(V)	33000
2	Highest System Voltage(V)	36000
3	No.of phase	3

4	Frequency(Hz)	50
1.03	**Variations**	
1	Voltage variation (%)	Plus/Minus 10%
2	Frequency variation C/S	48 to51
1.04	**Impulse Withstand Voltage (KVp/KVrms)**	(1.2 / 50 micro second wave)
1	HV	170
2	LV	Not applicable
1.05	**One minute power Frequency withstand voltage (KV)**	
1	HV	70
2	LV	3
1.06	**RATING**	
1	Rated voltage of L.V(V)	406
2	Max tem rise in winding (above 50 degree ambient)	55
3	max tem rise in winding (above 50 degree ambient 50'c)	50 (oil should be EHV grade)
1.07	**WINDING CONNECTION DETAILS**	
1	Connections	
a	H.V winding	Delta
b	L.V winding	Star
c	Neutral brought out for earthing	yes-but not to be earthed
2	Tapings	
a	No. Of positions	7- As per ALWAYE-Bangalore
b	Range	±7.5 in step of 2.50%
c	Voltage of each step	2.5% of rated voltage
3	Vector Group Three Wdg.	Dyn11
1.08	**Efficiency at unity PF**	**At 0.85 PF**
1	At 100% load	98.65%
2	At 75 % Load	98.88%
3	At 50% load	99.07%
1.09	**Termination details**	
1	HV/LV	
2	No. of Run on LV side	
1.10	**LOSSES AND OTHERS**	

1	No load losses at rated frequency & voltage	
2	CU losses at rated current and rated frequency at 75 degree c	
3	Percentage Impedance %	
4	Regulation at full load at 75 degree c unity power factor at 100% load	
5	Regulation at full load at 75 degree c .8 power factor at 100% load	
6	IS	
1.11	**Sheet Thickness**	
	Side Sheet	
	Top & Bottom sheet	
1.12	Harmonic Suppression/ (as per IS) for 3rd, 5th & 7th Harmonic	
1.13	Flux Density	
1.14	Tank to be pressure tested for mechanical strength.	
NOTE:- Transformer Marshalling box shall be SCADA complied and all related accessories will be a part of this specifications.		

;.5 Technical SPECIFICATION FOR LT PVC POWER & CONTROL CABLES

Suitable for continuous operation under the following tropical conditions-;.

1. Maximum ambient Temperature degree C) 50
2. Maximum temperature in shade (degree C) 45
3. Minimum temperature in air (degree C) in shade 3.5
4. Relative Humidity (%) 10 to 100
5. Maximum annual Rainfall (mm) 313
6. Maximum Wind Pressure (kg/mm2) 150
7. Maximum altitude above mean sea level (Meters) 1000
8. Isoceraunic level (days/year) 15
9. Seismic level (Horizontal acceleration) 0.3 g.
10. Ground temperature (deg. C) 40
11. Thermal Resistivity of soil (ohm. met) 50
12. Depth of laying for 1.1 kV (cm) 75 cables installed singly for twin / multi core
13. Cables and 3 Nos. of single core cables in Trefoil Touching. Also can be laid on trays over the ground.
14. Hot and dry climate,

Activity-6- Quality Assurance Plan

6.0 The quality assurance plan is the basic requirement of any project-this work starts right from the day, the concept and thought of designing and implementation of any project is taken up.

An experienced team of engineers must be constructed to look over the complete concept. This team will independently oversee and do the analysis of each and every activity like Design, Selection of Main Equipments, Selection of Contractors, Checks during construction at any stage as found necessary.

The design team/procurement team/site team/all other agencies, must give all feed back to the QAP team to Independently examine the scheme and give their feedback about the overall quality of the plant. A free exchange of views should take place between all the teams.

6.1 Construction Quality Management System-The following format is containing all activities that the construction engineers have to observe during the course of the construction of the plant in totality. The activity is listed here below, but the observations are to be predetermined by the site engineer, based on his own experience, manufacturers guidelines and fundamentals of the concerned engineering electrical/mechanical/ structural etc. The list is exhaustive, but not complete, as field engineers can find many other monitoring points and add to the list.

Sr. No.	**Equipment/Area/Work**	**Quality Engineer's Observation**
	Overall Plant Works	
1	Survey	
2	Boundary Wall	
3	Individual plant Location & Roads	
5	Store Management	
7	Piling	
8	Inverter Room	
9	Control Room if Applicable	
10	Foundation Work (DCJB, AJB, VCB, T/F)	
11	Structure / Module	
12	DC Work	
13	AC Work	
14	HT Yard	
15	Installation of components	
16	Transmission Line	
17	Electrical Connection	
21	Test Reports	
22	Pre Commissioning of plant	
23	Connectivity of plant	
24	Commissioning of each equipment	
26	Material Consumption	
27	Bills of Contractors	
28	Hand over / Take Over	
29	Punch points Compliance	
30	Project Closing	

6.2 HT Yard/Earthing/Equipment mounting and interconnections

Area/Field	Switchyard	
	Marking of Switchyard	
Civil Work	Marking of pole with all aspects	
	Excavation	
	Black Bitumen Painting of Grounded parts	
	Erection of poles/structures	
	Checking diagonally and by water Level	
	Excavation work for Earthing Electrode	
	Earth pit Chamber	
	Earth pit Cover	
	Installation of Meter Stand	
	Casting of Meter Stand	
Electrical Works	Fabrication of bracing, Channel Support	
	Fitting of DO Fuse	
	Fitting of isolator with Handle	
	Fitting of Earth switch with handle	
	Fitting of Fuse Wire	
	Fitting of Bus bar	
	LA fitting	
	Jumpring from LA to Busbar	
	Connection from bus bar to CT-PT	
	Connection from CT to HT cable	
	Earthing of Components	
	Three separate earthing of LA till Earth pit	
	Earth Mesh Completion of 4-pole	
	Cable clamp of HT Cable both side rising & lowering of 4-pole	
	Fitting of Meter Stand Box	
	Cable Routing	
	Termination of Metering Cable both side CT-PT & Meter Box	
	Insulated Cover of Isolator Handle	
	Auxiliary T / F Complete installation	
	MCB installation &Auxiliary Cable laying	
	Earthing of HT Cable	

	Smooth Operation of Isolator & Earth Switch by using Petroleum jelly etc.	
Miscellaneous	ETC of power and auxiliary transformers	
	Conduit work of Metering Cable	
	Fencing of Switch Yard	
	Leveling and Cleaning	
	Earthing Resistance value of Electrode in 4 Pole	

6.3 Inverter Room

1	Demarcation	
2	Excavation	
3	PCC	
4	Stone Masonry up to DPC level	
5	Cable Trench	
7	Brick Masonry up to Lintel beam	
8	Reinforcement of Lintel beam & Chajja	
9	Lintel beam &Chajja casting	
10	Brick Masonry up to Wall beam	
11	Wall beam & Roof Shuttering	
12	Wall beam & Roof Reinforcement	
13	Beam & Roof Casting	
14	Parapet wall	
15	Inverter Plinth fixing	
16	Ramp	
17	Plaster – Inside, outside wall & cable trench	
18	POP & color	
19	Curing	
20	Ducting Provision	
21	Checker plate, Window and door	
22	Exhaust fan for cooling of Inverter	
23	Internal wiring of inverter room	

6.4 Pilling Activity

1	Demarcation of location of each Pile	
2	Excavation	

3	Leg fixing and alignment	
4	Leg casting	
5	Measurement and projection check	
6	Curing	
7	Capping/Muffing	
8	Muffing paint	

Activity-7- DEVELOP INSTALLED COST ESTIMATES FOR THE SOLAR FARM

7.1 Start from the cost of land and the associated preliminary civil works.

7.2 Prepare a BOQ of all materials required for the power project.

7.3 Asses cost after contacting the best suppliers of the items.

7.4 Placed below is a sample consolidated Bill of Quantity of all materials required in any solar power project of any capacity. Only the quantity and specifications will vary and depend on the size of the plant in MW and equipment selected. Detailed specifications of main items are given separately.

Consolidated BOQ of any solar PV Power Project- Technical specifications given is indicative only. You have to prepare detailed technical specifications based on the equipments selected by you for your plant.						
Sl.No.	**Item**	**General Technical Specifications**	**UNIT**	**Qty.**	**Rate**	Cost
1	**Solar Field-DC area equipments**					
1.1	Module Mounting Structure	Suitably designed single/double portrait t bear the load of modules with a layer of two modules in a vertical column. Hot Dipped Galvanized Steel Structure for seasonal tilt of 5/13/26/43 deg. with horizontal. We can also go for triple portrait structure to save number of structures.	Kg.			
1.2	Solar Modules	Crystalline/Thin film-with specific mention that the modules are free from PID.	Nos.			
1.3	String Combiner Box	Outdoor type, IP 54, for 24 inputs, and one output, with fuses rating 10 amp. on both +ve and -ve, for 1000 volt DC, having Surge protection,	Nos.			

1.4	Inverters	...KW, self cooling, no forced air required, with input DC current as 1000 amp and D[illegible] input Voltage as 1000, giving AC output at.......Kv	Nos.			
1.5	MC4 Connectors	Rated voltage/Current 1000VDC,15Amp,Class-A for 2.5/4/6 sqmm PVC cable.	Nos.			
2	**Cables**					
	LT Cable					
A	4Sqmm Cu cable(DC) (Module to SCB)	1CX4 sqmm, 1.1KVstrandedcopper,un armored, suitable For outdoor mounting and exposed to hot direct sun light with anticipated temperature rise to 90 deg. cent.	KMtr			
B	150/240 Sqm[illegible] Al cable(DC) (SCB to Inverter)	1Rx2Cx150 sqmm, 1,1KV Aluminum, armored, suitable for outdoor mounting and underground laying.	Mtr			
C	3CX300Sqmm Aluminum cable(AC) (Inverter to Transformer)	Aluminum, stranded,1.1 kv, armored, XLPE insulated	Mtr			
D	3.5CX35Sqmm Cu cable(From Aux transf. to AC supply)	3.5CX35Sqmm Copper cable	Mtr			
E	Auxiliary supply cable	3CX2.5Sqmm	Mtr			
		3.5CX10Sqmm	Mtr			
	HT Cable					
F	HT power cable-3CX185 sqmm	33 kv, Aluminum, armored	Mtr			
	SCADA cables					
G	OFC	Multi-core	Mtr			
H	RS-485(Data cable)	4Core-0.5sqmm cable	Mtr			
3	**AC AREA Equipments**					
3.1	Power Transformer-	...KVA,33kv .,Three winding(tw[illegible] LV/one HV) ,oil filled, ONAN, off circuit tap changer,YnD-	Nos			

		11,copper wound complete with Bucholtz, OTI, WTI, PRV, MOG, wired Marshalling Box, with both side cable boxes.				
3.5	Plant Earthing and pits	Network of 50x6mm,25X3 GI flat, along with earth electrodes,2.5 mete deep with perforated pipe	Mtr.			
		GI Srtip-50x6 GI strip	Mtr			
		GI Srtip-25x3 GI strip	Mtr			
		3Mtr long earth electrode-dia 50MM	Nos			
		Charcoal, salt	KG			
3.6	33/.415 kv Au Transformer	25 kva,33/.415 volt, copper wound oil filled, off circuit tap changer, with cable terminatio on the LV side	Nos.			
3.7	415 volt MCB	415 volt, one input and5 outputs for 3-phase. Outdoor pole mounted, aluminum bus bar	Nos.			
4	Weather Station	Global irradiation sensor, temperature sensor, wind Speed and direction sensor, ambient temperature sensor, rain gauge sensor, relative humidity sensor along with loc direct reading and SCADA compliant card.	Set			
5	SCADA Systen		Set			
6	Computer station for SCADA viewin and Monitorir	Viewing and Monitoring of complete plant	Set			
7	**Miscellaneous Items**					
7.1	Ferrules	All relevant sizes	Nos			
7.2	**Cable Glands for-;**					
A	4sqmm cable	For both positive and negative cable at SCB point	Nos			

B	150Sqmm cable	For both positive and negative cable at SCB end and at inverter end.	Nos			
C	300Sqmm cable	At inverter end and transformer LT side	Nos			
D	185Sqmm cable	At Power transformer HT side/Auxiliary TF	Nos			
7.3	**Lugs for Cable Sizes**					
A	4sqmm cable	For both positive and negative cable at SCB point	Nos			
B	150Sqmm cable	For both positive and negative cable at SCB end and at inverter end.	Nos			
C	300Sqmm cable	At inverter end and transformer LT side	Nos			
D	185Sqmm cable-HT	HT side	Nos			
7.4	**Cable Trays**					
A	75*50*2MM	GI hot dipped galvanized ladder/perforated cable tray with cover	Mtr			
7.5	**Termination Kit for 33KV**					
A	185Sqmm cable-HT	Indoor type	Nos			
B	185Sqmm cable-HT	Outdoor type	Nos			
7.6	**Cable Tie**	All relevant sizes	Lot			
7.7	Cable identification tag	All relevant sizes	Lot			
7.8	Fire fighting arrangement for switchyard and for inverter building with sand buckets	All relevant types and quantity	set			
7.9	Plant Lighting	20 Watt solar lighting as per past purchase specs.	Nos			

8	**33kv metering set at plant end and GSS end.**	ABT meter Complete with CT ,PT, meter and associated accessories as per DISCOM norms (20/5-5Amps,33KV/.110KV -PT)	**Nos**			
9	**Four Pole Structure**					
1	I-Beam	Four Pole support I-Beams 175x100x6MM,8Mtr long	Nos			
2	MS channel	100x50x6MM ISMC for suppor and bracing	Mtr			
3	Cable pipe for CT-PT cable	As per requirement	Nos			
4	Disc Insulator(each disc of 11kv rating)	Disc insulator with string hardware fitting set for 33kv (three disc)suitabl for dog/panther conductor (bolted/compressio type)	Set			
5	ACSR dog/panther conductor	As per requirement	Mtr			
6	danger Plate-33kv type	As per requirement	Nos			
7	M.S.Nut bolts with washer and clamps assorted siz	As per requirement	Nos			
8	Fencing of S/S land with barbed wire and supports	As per requirement	Nos			
9	33kv T-clamp	As per requirement	Nos			
10	33kv HG fuse/ Horn gap	As per requirement	Set			
11	33kv,630Amp isolator	As per requirement	Set			
12	33kv LA statio class type wit clamp	As per requirement	Set			
13	Auxiliary transformer-	Rating-25KVA,33KV/.440KV	Nos			

14	CT-PT cable	As per requirement	Mtr			

Activity-8 Explanations on the terms-; Solar Cable, Maximum Power Point Tracking, Master Slave Configuration, Potentially Induced Degradation, PRACTICAL Test Conditions and Normally Operating Cell Temperatures.The following terms have a significant impact on the overall design and cost of a PV-Solar Project these are:-

- ✓ **Solar Cable,**
- ✓ **Maximum Power Point Tracking,**
- ✓ **Master Slave Configuration,**
- ✓ **Potentially Induced Degradation,**
- ✓ **PRACTICAL Test Conditions and Normally Operating Cell Temperatures**

These five terms are the talking points for all, engineers-designers-investors-EPC people-who are involved in the solar industry. If these points are understood by the designers, there will be a vast effect on the techno-commercial aspect of the Power Project. Actually these factors do have technical impact on the design, but there are not much of feedbacks/ studies to emphasize that these are real issues. A lot is yet to come on these aspects.

We, here below try to explain each of these points in detail.

8.1 Solar Cable: -

In electrical engineering there is nothing like a solar cable. The cables are designed and manufactured to certain PRACTICAL specification (ISS in India and IEC in Europe). TUV certified or TUV compliant cables, as mentioned by cable suppliers needs thorough understanding of the issue.

1. For every cable there are fixed parameters like-Voltage Rating, current carrying capacity, unit length resistance, temperature withstand value type of insulation etc.

2. The most important is the temperature- every cable has to withstand a temperature of 90°C during loading. This 90 C temp. rise, with ambient of 50°C and at full capacity loading of the cable, is never achieved. Practically, this goes to a maximum of about 70C. So reputed make cables do not face any damage or degradation-neither in the conductor nor in the insulation.

To further counteract the problem of temperature rise, we provide higher current carrying capacity (CCC) cables for every section. For examples, for string formation, the maximum current is 8.5 Amps, but we select 4 Sqmm. Copper cables having CCC as 45 Amps, which is 11 times higher than actual current loading in each string. So the I^2R losses, (which also increase the temperature of cables) are negligible.

With selection of higher capacity cable, the short circuit requirement of the cable is automatically addressed.

The only criteria for selection of cable should be the quality of copper and Insulation material and also a reputed make cable should be selected. The good manufacturers in India cater to these criteria. In fact Indian make cables are being exported in large quantity. So shifting to indigenous 4 sqmm cable of Indian make can be done, instead of going with so called imported solar cable. A heavy cost can be saved.

8.2 Maximum Power Point Tracking:-

Function of a MPPT is analogous to the gear assembly in a car.

- First gear corresponds to 1 kmph,

- 2nd gear corresponds to 20 kmph,
- 3rd gear corresponds to 40 kmph,
- 4th gear corresponds to 60 kmph.

When the gear is in the wrong gear position, the wheels do not receive maximum power delivered by the engine. That's because the engine (The power generator/solar modules) is running either slower or faster than its ideal speed range. The purpose of the gear assembly (MPPT in Inverter) is to couple the engine to the wheels, in a way that lets the engine run in a favorable speed range (maximum power output) in spite of varying acceleration and terrain (soiling/shading/solar radiation/module mismatch/string output mismatch etc.).
The Inverter has a facility to continuously track the input D.C.Power parameters
The DC to A.C. inversion process is linked to the maximum D.C. power available on the D.C. Bus of the inverter; therefore the inverter tracks and uses maximum input power from the field and converts it to AC power, hence the name -MPPT.

But the total system, starting from Module-To String-To S.C.B. - to Inverter, itself delivers a stable-but continuous power to the Inverter; therefore the inverter does not have much to track continuously on the stable sun insolation day. The power input curve at inverter input will be, almost a straight line, depending on the sun radiation.

- The soiled modules give an output-though low-but stable- depending upon their condition.
- Module miss- match loss is neutralized in the string-again it gives a stable output voltage.
- Sun shines equally on all the modules at one place, hence again a stable output from each module is given out during whole day.
- The string miss match is taken care of in the SCB and a stable output is released to Inverter.
- The output of all S.C.B. meets at the D.C. Bus of Inverter therefore again an adjusted minimum, but almost stable output is delivered to the Inverter Bus.
- Yes, power delivered will be low at 9AM and 5PM, and minimum at 12:00 Noon.
- Therefore when the DC input power is almost stable in terms of voltage level, where is the need of MPPT.

Result:- As the power reaching the inverter is in an almost stable state, at particular time of the day, the MPPT system of the inverter does not have much to track and act at the point of maximum power input. Practically we observe that the variation/hunting in DC current reaching the inverter varies by about 1 to 2 amps, and the voltage remains almost constant.

But in small, battery storage solar plants, having one or two strings of PV modules, as in roof top/domestic plants, the effect of MPPT is very high.
People may differ from my observations and their views are welcome.

8.3 Master Slave Configurations of Inverters:-

The Inverters of higher rating (above 630 kVA) are modular type. Because of temperature rise limitations we cannot have bigger than 630 kVA size inverters. Depending upon desired capacity of inverter, a number of smaller size inverters (modules) are connected in such a way that the total capacity of the inverter increase to desired value.

Practically the inverter modules, so connected, are coupled in such a way that, the first inverter module in circuit starts getting loaded at the start. The moment the capacity of this first module is fully loaded, the second module starts getting loaded and the phenomenon repeats for other module in line. We therefore say that the first module works as a master and next works as a slave.

The analogy can be-number of overhead water tanks joined at the top with each other. If we fill water in the tanks from one source connected at the top of first tank, this first tank will fill first. Once it is full the next tank in series will start filling so on and so forth. So the first tank works as a master and the remaining tanks work as slaves.

Therefore this arrangement of modular inverters, joined together to form a big inverter, cannot have MPPT arrangement/facility. Do we need the MPPT and stick to a number of inverters of rating maximum 630 KVA for MW capacity power plants is the question-and my answer is NO. We can have large size of inverters made up of smaller modular inverters and forget about the advantage of MPPT. People may differ.

8.4 Degradation of Photo Voltaic Modules

Researches are being carried by the solar module manufacturers to study the reasons for degradation

of solar modules with time. The study started in 2005 when it was observed that some cells/modules got damaged during use at site. The phenomenon was continuously studied and the following facts were observed-

1. The string output of some strings got reduced.

2. It did not increase even when modules were cleaned and the terminations were tightened.

3. A continuous watch revealed that-;
- The module corners changed their color first to white spot to brown/black
- The output slowly started reducing and reached to zero.
-Voc also reduced slowly to zero.
-Surprisingly, this phenomenon was observed in only some modules of the same lot of manufacture.
-The reasons, such as improper handling and storage, water entering in the Modules were crosschecked, but this was not there.
-The phenomenon continued randomly and in some modules only.
-The system, to which the module was connected, was examined and it was detected that, there is a charge concentration near the module frame, which indicated that there is stray charge leakage from the module.
- This stray charge was travelling to the ground from the provided earthing system.

-The charge flow from negative to positive was analyzed, and it was revealed that there is ion flow with the charge. The question is whether this ion flow is the culprit for reducing the life of the module.

1. Finally the research zeroed to the term POTENTIALLY INDUCED DEGRADATION.

In 2005, Sun Power reported that their modules (N-Type Cells) experienced degradation with time. They searched for the reasons of blackening of the solar modules. These blackened modules slowly reduced their output power and finally failed.

Their research indicated that the reason for this phenomenon is potentially Induced Degradation (PID) of the solar modules. They also observed leakage of charge in the vicinity of the modules.

As the phenomenon is time related, so experiments were carried out:-
It was observed that:-

- The + ve grounded systems indicated PID, in N-Type Cells.
- The – ve grounded system indicated PID in P-Type Cells.
- The voltage gradient between PV circuit and ground (1000volt) caused leakage current.
- Leakage current level is determined by the insulation electrical conductivity that varies greatly with temperature and humidity.
- With +ve grounded system, the negative charge accumulate on the top surface.
- The charged surface will attract the electrons from the cell, forming the leakage current and destroying the p-n junction in the module.
- High humidity, high system voltage and high temperatures accelerated this problem. So it was established that there are four suspects who cause the PID, which are-Voltage, Humidity, Temperature and Grounding. The levels of these are:-

- Voltage – 600 and above with negative biased.
- Temp – 85'C and above
- Humidity – 70% and above
- Grounding – Both + ve and/or – ve is a factor but research is still going on to Confirm this fact.

1. **There is no International or Indian PRACTICAL on PID yet.**
2. Product qualification PRACTICALs IEC61215, IEC61730 and UL1703 do not evaluate modules 'Capability of anti-PID'.
3. Different PID evaluation programs are being carried out, but none of them has been verified to be effective so far.
    - The reason is, till the year 2014 there is no proven study available, which indicated that the phenomenon of PID is there and it follows certain PRACTICALs and conditions.
    - It was also observed that module deterioration did not follow an understandable system.
    - Modules, prepared with same material, in same lot, used in same location, and same atmospheric conditions, behaved in different manner. A very small percentage of such modules got deteriorated randomly, but rest continued to work satisfactorily.

- Quartz glass is very expensive, so it's not practical to use it in large scale

The different module suppliers have different views and intimate the end users, according to their research- some ask for grounding the negative and some ask for floating system of DC.

8.5 Standard Test Conditions and Normally Operating Cell Temperatures

Every electrical appliance/equipment has an internal electrical resistance. Under charged or live condition, this internal resistance offers resistance to the current that flows through the equipment, there by causing ohmic losses inside it. This loss appears as temperature in the equipment and its' temperature rises. Based on this raised temperature level, the output of the equipment finally depends.

But the equipment suppliers mention the output level of their equipments on the STANDARD TEST CONDITIONS (STC) which are pre-decided conditions of ambient temperature, solar irradiance and wind mass velocity. For inverters the STC temp. is 25 deg. cel. Irradiance is 1000 W/M sq and wind mass flow taken as 1.5 m/sec.

Therefore, selecting any equipment, whose capacity is tested on/specified on/based on STC conditions, is not a correct method. For example a 300 watt solar module will give only an output of 270 watt, because the value 300 is based on STC, whereas the value 270 is based on actual site conditions. The main difference in the STC conditions and actual site conditions is the temperature at which module is operating. In STC this temperature is taken as 25 deg.centi., where as, at site this temperature is 46 deg.centi. in Indian conditions most of the times.

This fact then introduces another term **NORMALLY OPERATING CELL TEMPERATURE (NOCT).** In Indian climate, the day time- sun available-temperature is varies vastly from 25 deg.cent. to 48 deg. Cent. Add to it, the on load internal loss related temperature of the live equipment, which is 20 deg.centi. Therefore for solar cells (and also inverters and transformers), we have to also consider this NOCT level for deciding their capacity.

Why we decide and calculate the number of modules in a plant on the STC based capacity (as indicated by the manufacturers and suppliers in their brochures) and not the NOCT based actual capacity, is very difficult to explain. However this is discussed here below in short, leaving many aspects for the designers to ponder upon

The reason is not technical-it is purely commercial. But even these commercial implications are very difficult to understand. Select the solar modules on STC-you reduce their number and save capital cost-but then by doing this-you reduce the generated energy-again a commercial loss. Compare the two losses and decide what is good for you.

Experts may differ from my above observations and their corrections/comments/guidelines are welcome. The topic of designing on STC or NOCT is open for discussion. But please note that in South Indian Area the STC conditions of 25/1000/1.5 is created between the months September to March every year. These months are called the clean months. Sky is washed by the rains, sun rays do not face any pollution, sun is bright but not that hot and wind mass disturbance is low.

Activity-9 Work's Planning-Site works are divided into three parts-; DC Area, AC Area and Civil Works.

It becomes a duty of the designer to assess the quantum of work for erection testing and commissioning of the complete plant. Therefore first of all we identify the total work, bifurcate it into small units and then make complete tender for the contractors to quote their rates.

9.1 Works of DC area- cabling, string formation and inverter installations-;Values filled are indicative only-

	1. The following are the works of DC area cabling, string formation and inverter installation alongwith their erection testing and commissioning.. **2. This is after the site is cleared from the civil engineering point of view.**	
	Module's accessories and installation.	Quantity
3.	**String formation –**Interconnection of 4 sqmm/6sqmm solar cable between modules by laying along the structure. Cabling dressing by cable tie etc. This work shall include termination of cable at both ends.	
3.1	**Laying of ICx4sqmm/6sqmm cable** between strings and string combiner box by dressing along the structure, laying inside the conduit pipe along the ground, sealing of cable entry by cable sealing foam. The work shall include laying of conduit pipe straight run, bends etc. Contractor shall perform insulation resistance measurement of all cables before and after laying the cable to assert the correctness of the cable.	

	Testing and continuity of each run of cable In presence of client/customer representative as per PRACTICAL engineering practice.	
3.2	**Termination and connection** by crimping lug and fixing of MC4 connectors, ferruling of string number at module end and SCB end for 1Cx4sqmm/6sqmm solar cable.	
4.	**Laying of cables**	
4.1	**Laying of armored cable 95/120/150**sqmm cable from SCB to inverter inside cable trench of Control Room. Contractor shall perform insulation resistance testing and continuity testing of each run of power cable in presence of client/customer representative as per PRACTICAL engineering practice.	
4.2	**End termination and connection** by crimping lug, glanding, and heat shrink sleeve and feruling of armored power cable at SCB terminal. Supply and installation of support arrangement for cable. It shall be inclusive of minor activities like cutting, drilling and fixing of gland plate etc.	
4.3	**End termination and connection** by crimping lug, glanding, and heat shrink sleeve and feruling of armored cable at inverter terminal. Supply and installation of support arrangement for cable. It shall be inclusive of minor activities like cutting, drilling and fixing of gland plate etc.	
5.	**Installation, Testing of string combiner box.** Installation and grounding of SCB on concrete slab/angle iron /any other method as per site engineers decision at site. Testing of SCB (i.e. testing voltage and current of each of the strings).Connection of Earthing for SCB, etc at ends.	
6.	**Installation, Testing of inverter**. Installation and grounding of inverter. Connection of Earthing for inverter. Associated power and control cabling, their termination and commissioning.	
7.	**Installation of earthing mat GI earth strip** directly buried type below ground level. This shall be inclusive of termination with earth pits, equipments, structure etc. This shall be inclusive of works such as bending, welding, cutting, painting, etc. Providing earth pits.	
8.	**Structure shall be body earthed** and interconnected. As per earthing design, each leg of the structure support and all equipments will be connected to the earth mesh with25X3 sqmm GI flat. Testing of earth value at the terminal of structure.	
9.	**All DC cables are laying in trench with all materials (Bricks and Sand if ordered as per drawing.**	
10.	**Trenching/digging and backfilling** properly after completion of the cable laying work. The backfilling will be done only after all cables are laid so that we may not require re-digging for any other cable. Contractor will see that cables laid in the trench and are still lying in open are not damaged by trucks and tractors. He will make proper pathway with the help of Hume pipe	

9.2 Works of Civil and AC area– cabling, four pole erection, metering, VCB installation, transformer ETC and protection settings

			Unit	Quantity
1	**Module and Mounting Structure – Civil**	Module and Module Mounting Structure Installation (Per MW)		
2	**Piling Civil**	Piling Civil Work (Nos.)		
		Cement For Piling (Bag)		
		Paint-Pile Cap (Bucket)		
3	**SCB, Inverter and Inverter Room Civil**	Inverter (Nos.)		
		Construction of room (Sq. Ft)		
		Trench Cover in Inverter Rooms		
		SCB (Nos.)erection		
		Paint-SCB Foundation		
		Cement for SCB Foundation		
		Window Filter		
4	**AC Area works**	Transformer ETC (Nos.)		
		Total plant earthing for both AC/DC area		
		AC Cabling Work (Per MW)		
		Switchyard Civil Work (Nos.)		
		Cement for Switchyard Foundation		
		Paint-Earthing Chamber Box		
		Paint-Switchyard foundation		
		Cable-Inverter To Transformer		
		Cable-Transformer To 4 Pole		
		Cable- 4 Pole To TL		
		Cable-Auxiliary cable for Transformer		
		Cable-Metering		
		Cable Trays/Conduits		
5	**4 Pole installation**	CT-PT, Meter		
		Complete four pole erection- civil works		
		Auxiliary Transformer		
		Disc Insulator		
		Channel-100X50		
		Channel-125X65		
		Channel-I Beam		
		Horn Gap fuse		

		Isolator with Earth		
		LA Station Class type		
		T Clamp		
		PG Clamp		
		Red Oxide		
		Paint-4 Pole		
		Nut Bolt-4 Pole		
		4 Pole Installation		
6	**Aesthetics and Security**	Gate		
		Boundary Wall		
		Security Guard Cabin		
		CCTV		
		Street Lights		
		Danger Plates		
		Light and Fan for Inverter Room		
		Roads		
		Street Light Installation		
		Finishing		
7	**SCADA supply and Installation**	RTU		
		Isolation Transformer		
		Cable-Communication		
		Cable-Power Cable for SCADA		
		SCADA installation		

With above identified works, we can estimate the cost of the projects also.

Activity-**10 CONDUCT ECONOMIC ANALYSIS OF THE PROPOSED SOLAR FARM'S CHARACTERISTICS**

This work is of highly commercial in nature, and hence is to be done by the internal confidential team of the organization. The works involved are highlighted here below. There are many more issues in this commercial work, and needs involvement of the accounts experts-;

MW PV System Annual Energy Generation: 16, 50.000 kWh
Life of PV System: 25.0 years
Annual PV System Output Degradation: 1%
Power Purchase Agreement Period: 25.0 years
Investment Tax Credit (ITC): 30% of system capital cost accruing in Year 1 of operation
Depreciation Method: 5-year Modified Accelerated Cost Recovery System (MACRS)
Finance Period for 3rd Party Developer: 10 years
Finance Interest Rate: 13%
Amount of Financing required: 100% of capital cost

Annual Energy Escalation Rate for Conventional Utility Electricity Purchases: 3.5%
Annual Operations and Maintenance Cost of Solar Plant Rs.1.00/kWh

Activity 11: Prepare a final report and the detailed drawings-SLD, Cabling, String formation and all civil drawings.

11.1 Once the plant land and its capacity are decided, we have to proceed with detailed designing of the project. For this, the string formation, structures required details, SCB nos. inverter capacity, transformer type and capacity is to be calculated.

11.2 Drawings- The following drawings of projects have to be prepared.

1. AC SLD
2. Cable layout, scb position and associated cabling
3. Earthing and lightening protections
4. Four pole structure
5. GA Layout
6. Civil Drawings of -; Transformer foundation, Inverter Room, VCB foundation, SCB foundation, Earth pit, Control Room with sewerage.
7. Piling Layout
8. String Arrangement
9. Cable trenches/tray arrangement

The following drawings of a 1-MW actual plant are as given below. These drawings are self explanatory, more so after reading the above technical thesis of the complete solar PV power plant.

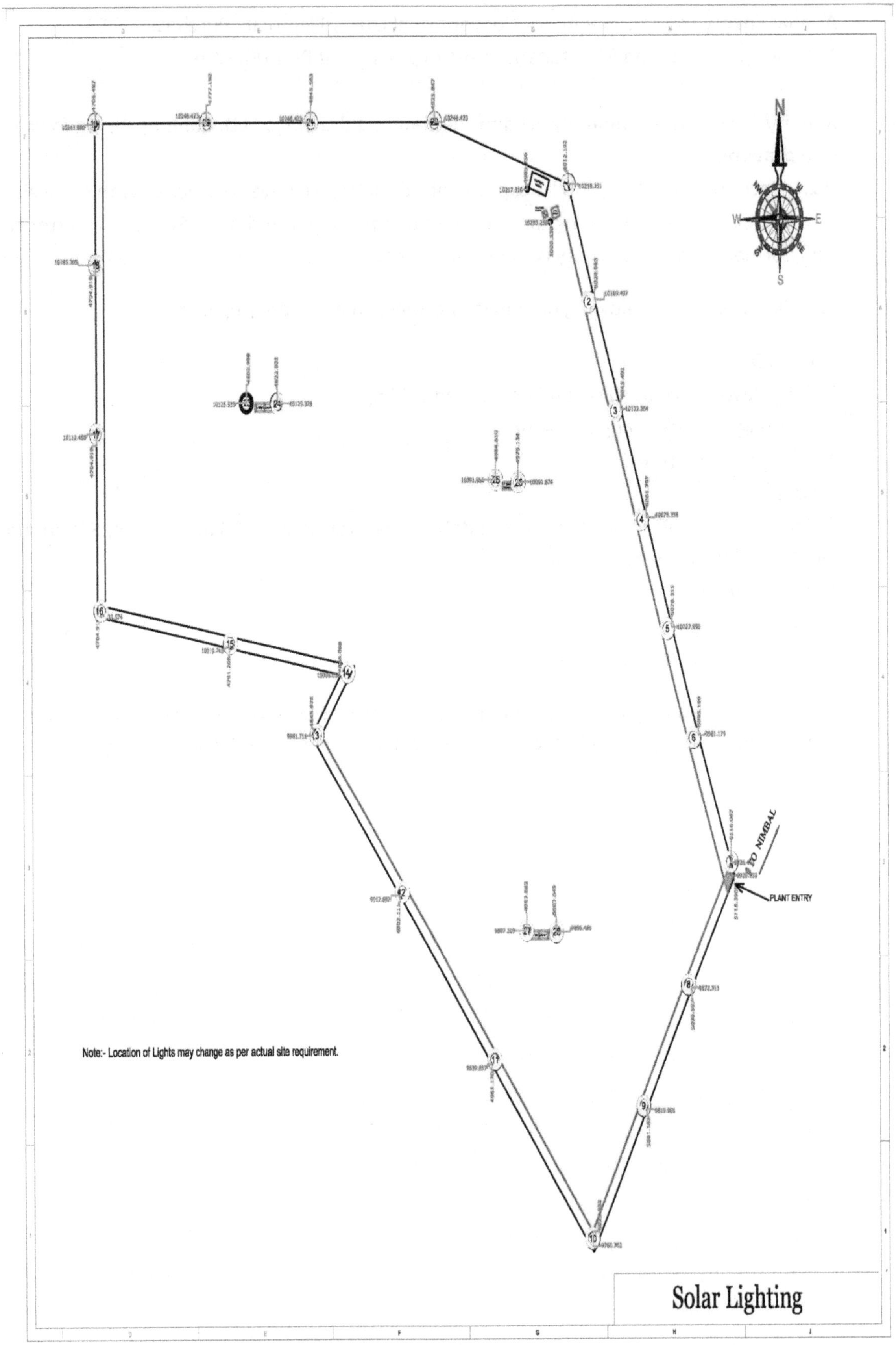
N
W
E
S
TO NIMBAL
PLANT ENTRY
Note:- Location of Lights may change as per actual site requirement.
Solar Lighting

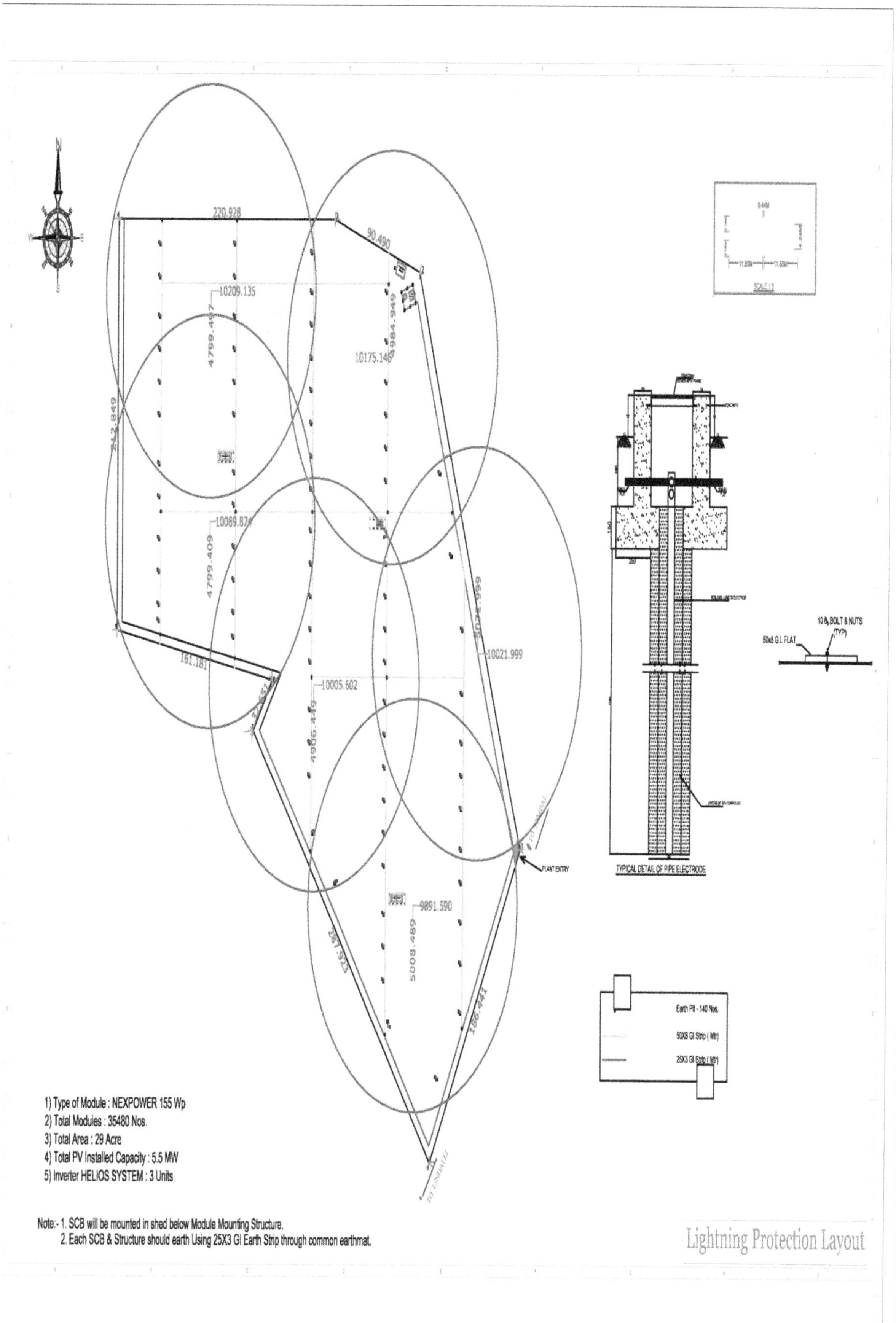

220.928
90.490
10209.135
4799.497
10175.146
4984.949
242.849
10089.874
4799.409
161.181
10005.602
4906.449
10021.999
9891.590
5008.489
281.525
186.441
PLANT ENTRY
TYPICAL DETAIL OF PIPE ELECTRODE
10 ϕ BOLT & NUTS (TYP)
50x8 G.I. FLAT
Earth Pit - 140 Nos.
50X8 GI Strip (Mtr)
25X3 GI Strip (Mtr)
1) Type of Module : NEXPOWER 155 Wp
2) Total Modules : 35480 Nos.
3) Total Area : 29 Acre
4) Total PV Installed Capacity : 5.5 MW
5) Inverter HELIOS SYSTEM : 3 Units
Note:- 1. SCB will be mounted in shed below Module Mounting Structure.
2. Each SCB & Structure should earth Using 25X3 GI Earth Strip through common earthmat.
Lightning Protection Layout

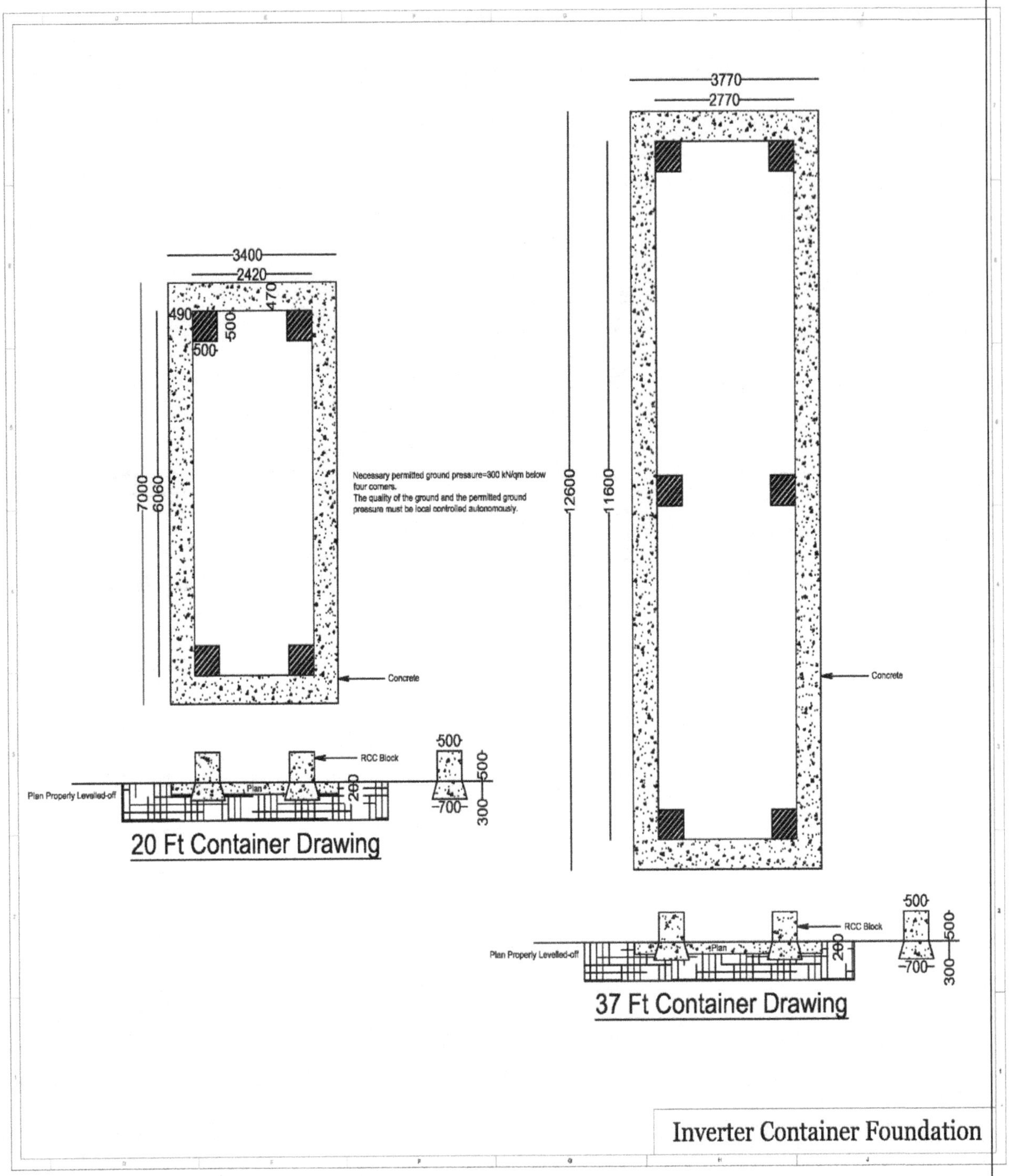
3400
2420
470
490
500
500
7000
6060
Necessary permitted ground pressure=300 kN/qm below four corners.
The quality of the ground and the permitted ground pressure must be local controlled autonomously.
Concrete
RCC Block
Plan Properly Levelled-off
Plan
200
500
500
700
300
20 Ft Container Drawing
3770
2770
12600
11600
Concrete
RCC Block
Plan Properly Levelled-off
Plan
200
500
500
700
300
37 Ft Container Drawing
Inverter Container Foundation

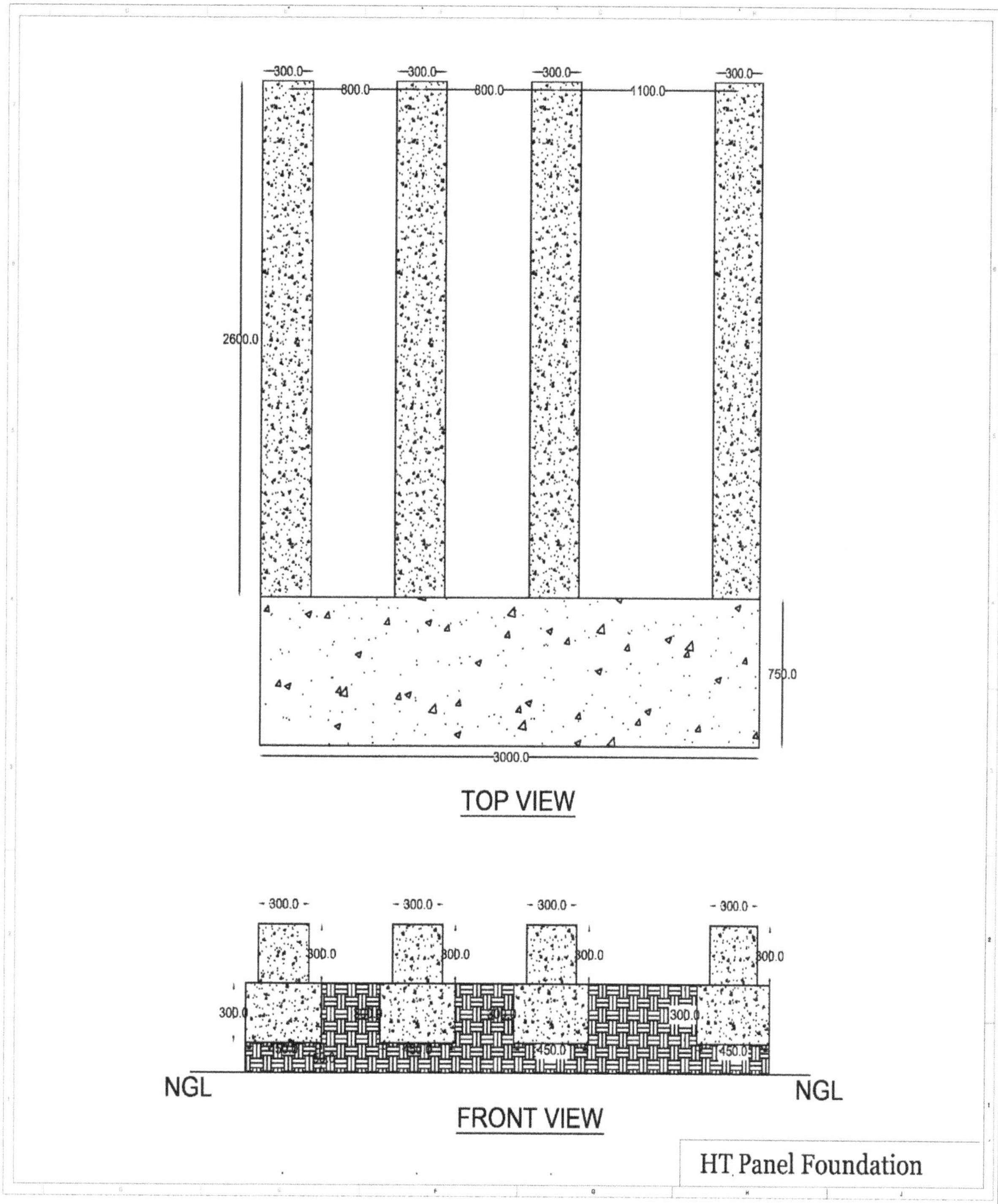
300.0
800.0
1100.0
2600.0
750.0
3000.0
TOP VIEW
300.0
450.0
NGL
FRONT VIEW
HT Panel Foundation

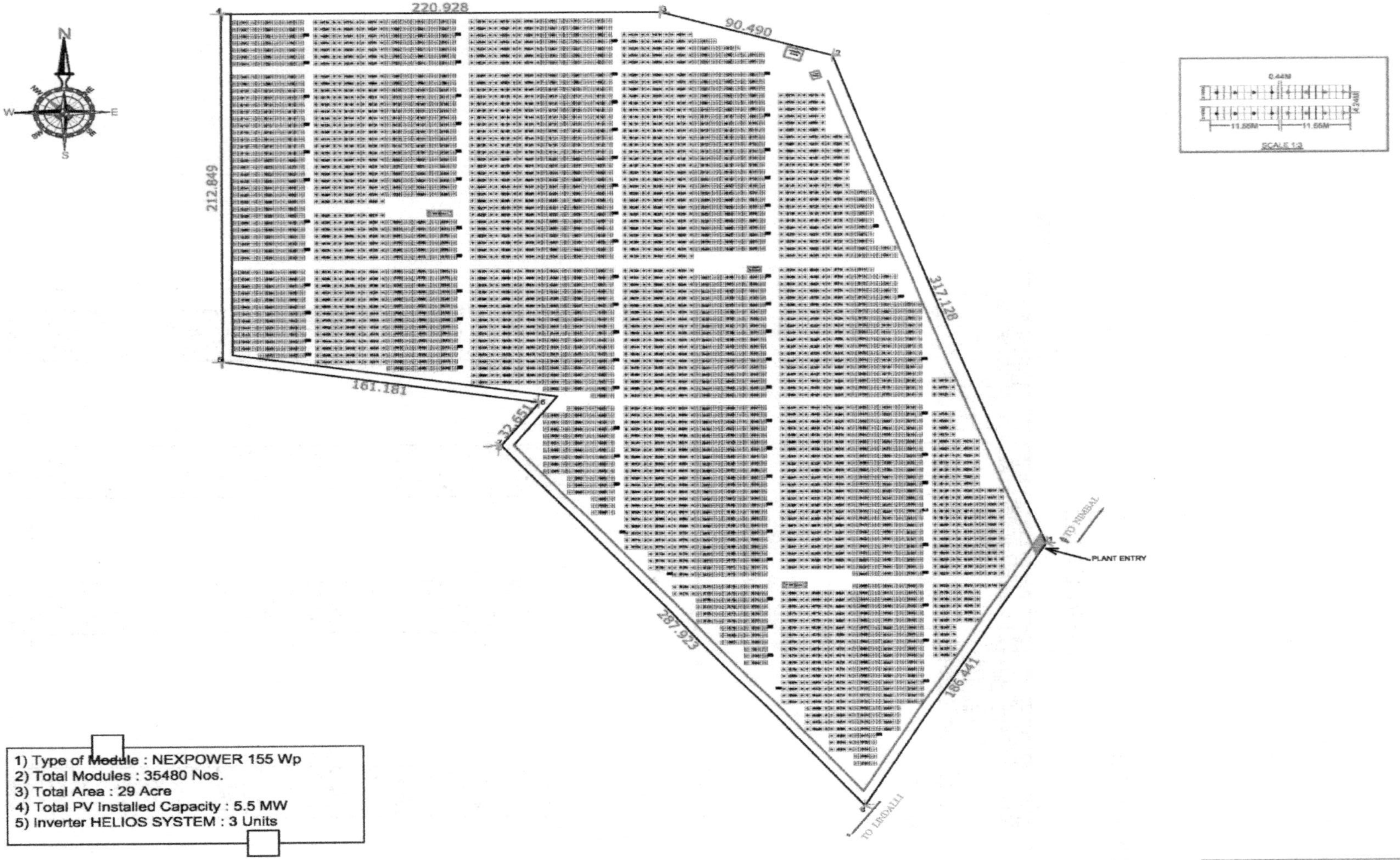

1) Type of Module : NEXPOWER 155 Wp
2) Total Modules : 35480 Nos.
3) Total Area : 29 Acre
4) Total PV Installed Capacity : 5.5 MW
5) Inverter HELIOS SYSTEM : 3 Units

General Arrangement Layout

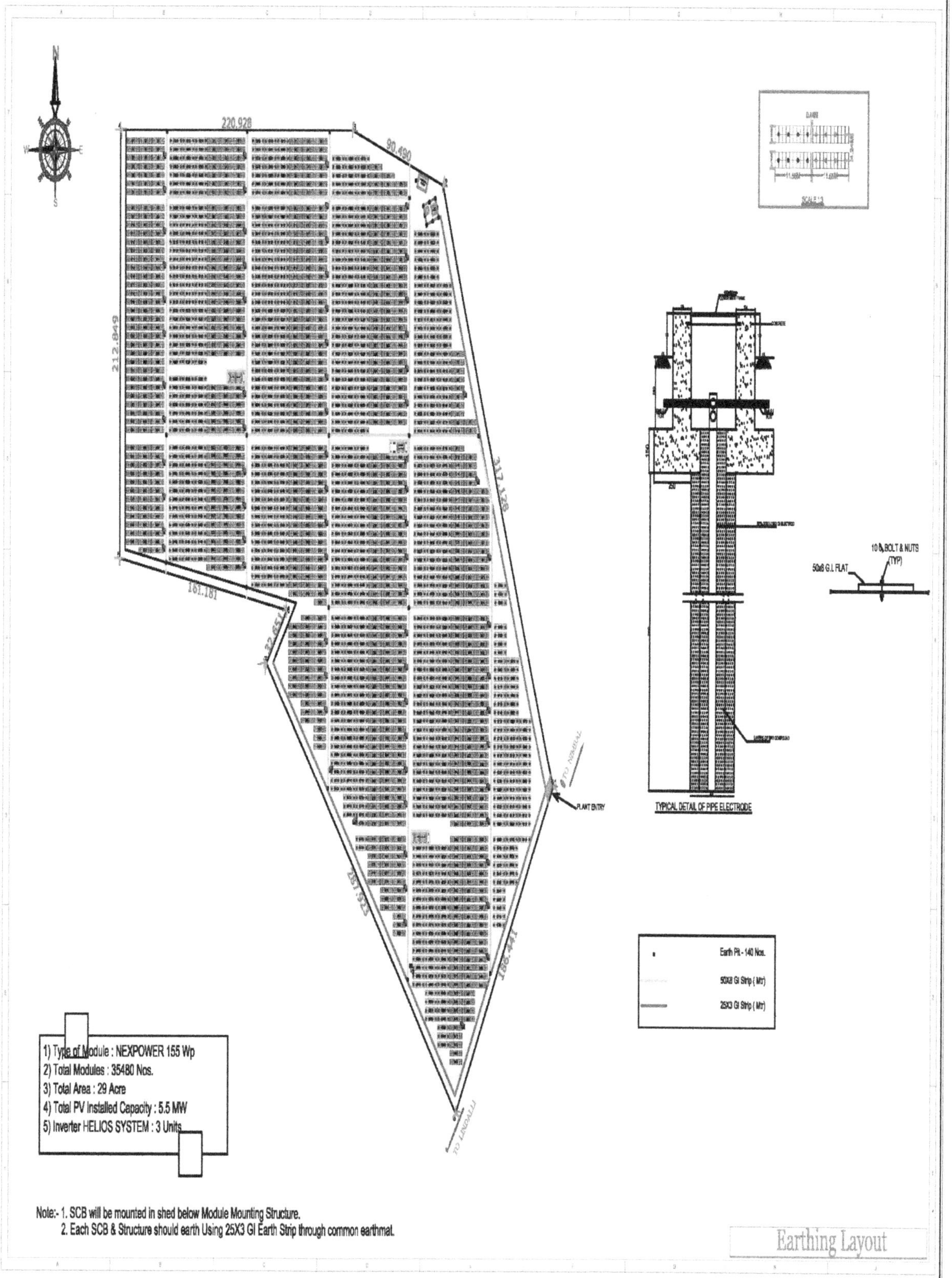

N
220.928
90.490
212.849
317.128
161.181
32.651
281.923
186.441
PLANT ENTRY
TYPICAL DETAIL OF PIPE ELECTRODE
10 Ø BOLT & NUTS (TYP)
50x6 G.I. FLAT
Earth Pit - 140 Nos.
50X6 GI Strip (Mtr)
25X3 GI Strip (Mtr)
1) Type of Module : NEXPOWER 155 Wp
2) Total Modules : 35480 Nos.
3) Total Area : 29 Acre
4) Total PV Installed Capacity : 5.5 MW
5) Inverter HELIOS SYSTEM : 3 Units
Note:- 1. SCB will be mounted in shed below Module Mounting Structure.
2. Each SCB & Structure should earth Using 25X3 GI Earth Strip through common earthmat.
Earthing Layout

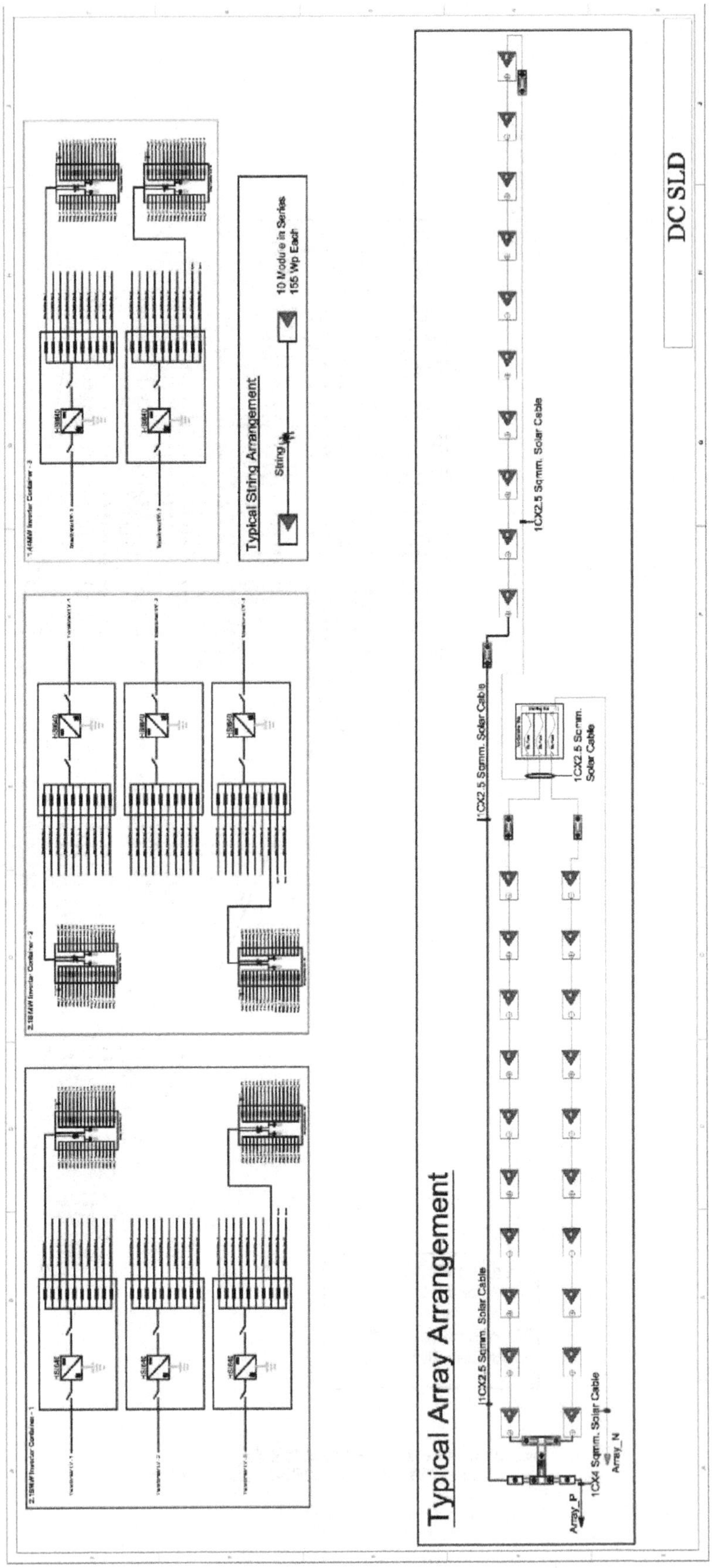
DC SLD
Typical Array Arrangement
Typical String Arrangement
String
10 Module in Series
155 Wp Each
1CX2.5 Sqmm. Solar Cable
1CX4 Sqmm. Solar Cable
Array_P
Array_N

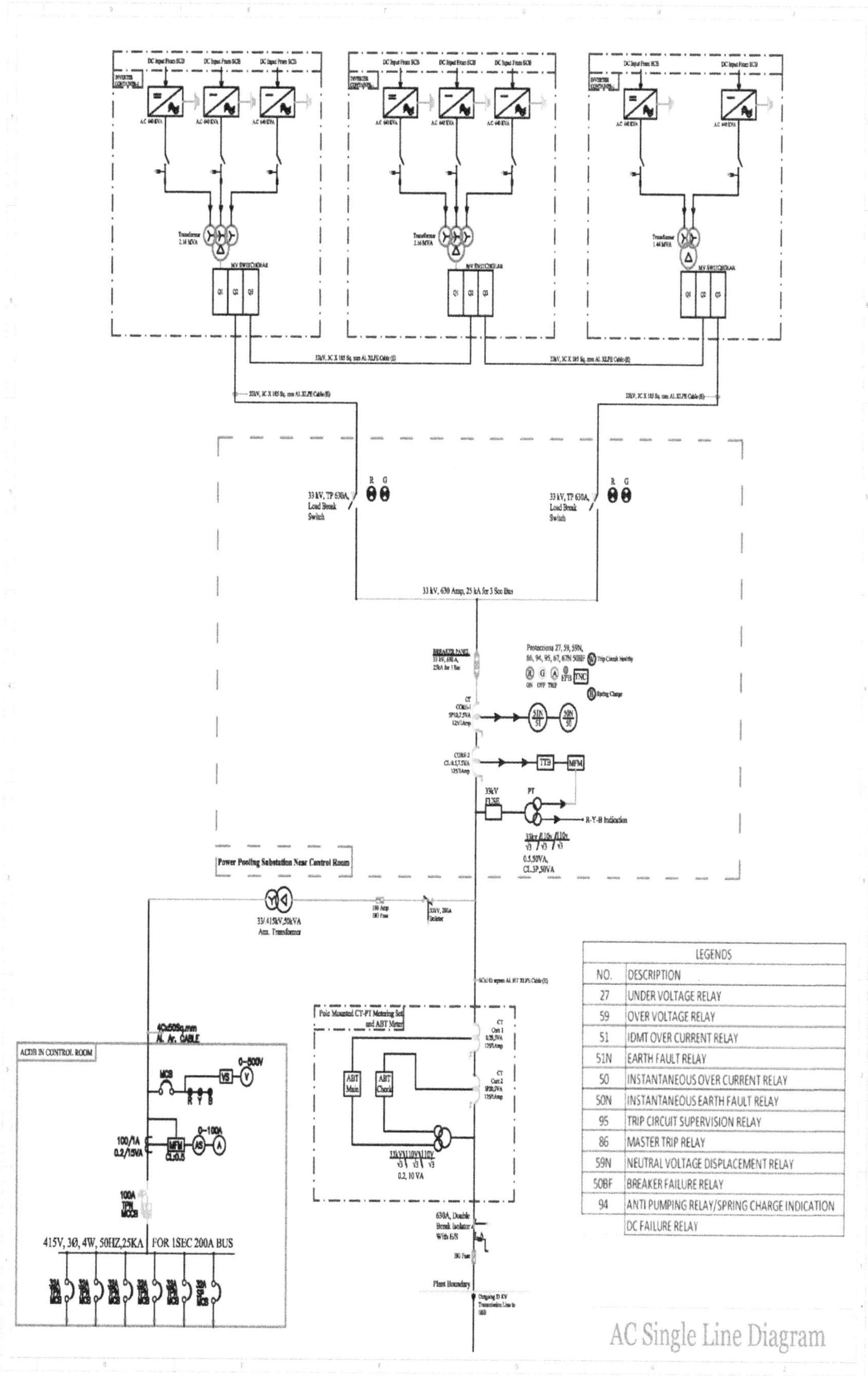

LEGENDS	
NO.	DESCRIPTION
27	UNDER VOLTAGE RELAY
59	OVER VOLTAGE RELAY
51	IDMT OVER CURRENT RELAY
51N	EARTH FAULT RELAY
50	INSTANTANEOUS OVER CURRENT RELAY
50N	INSTANTANEOUS EARTH FAULT RELAY
95	TRIP CIRCUIT SUPERVISION RELAY
86	MASTER TRIP RELAY
59N	NEUTRAL VOLTAGE DISPLACEMENT RELAY
50BF	BREAKER FAILURE RELAY
94	ANTI PUMPING RELAY/SPRING CHARGE INDICATION
	DC FAILURE RELAY

AC Single Line Diagram

The string formation consumes a lot of the solar cable and the selection of the above detailed combinations has another consideration, that is, the reduction in total length of this costly solar cable. If we are able to reduce the solar cable length by selecting proper module placing concept, minimum gaps between the structures, we save a lot of money in the cable cost and also the related energy loss due to I square R losses.

A study of these drawings will indicate that, in these designs efforts have to be made to-;

- Reduce the losses,
- Reduce the overall cost,
- Increase Sun Window hours each day to increase generation,
- Reduce HT cabling by going into the overhead transmission line concept to reduce cost, losses and faults.
- Provide cable trays for laying the cables cross-country to reduce cable length-thereby reducing losses and avoid all trench works.
- Further, in particular, we have introduced the seasonal tilt, four times in a year on predefined fixed dates. The angles will be-5/13/26/43 degree during summer to winter. This will increase our generation level.
- The earth mesh design has been changed to make it more effective with less cost.
- We have introduced the inbuilt water washing concept, so that it is done without the tractor
- mounted water washing, which creates lots of problems at site.
-

11.3 Basic design data required to prepare the GA layout

Study the data provided by the suppliers of -;

- Modules
- Inverter
- Transformer.
- The datasheet is given on STC (PRACTICAL Test Condition), which are 1000w/msq of radiation, at 25 degree celsius. with air mass as 1.5.
- This STC never reaches in actual site conditions-therefore designing the PV system in totality on STC conditions is open to discussions.
- Use of Voc is however necessary, because in case of string open circuiting, the condition of Voc can be created.
- While designing the plant in totality, we will consider the Nominal power of the module at the outset. Based on this nominal power, the total no of modules will be calculated.
- But for string formation and deciding the number of strings per SCB, structure quantity, module placement etc. we will consider the Voc of the module as given in the module sheet.
- For deciding the capacity of the inverter, we consider Vmp and Isc.
- Every inverter supplier gives a range of inverter capacity, for both DC side and AC side. Of course for short time, temperature rise limitations permitting, overloading of the inverters is permitted. The permitted overloading is a bonus only and not deciding criteria.
- Same is true for the transformer. Transformer is a static device and it can be loaded to +10% continuously over and above its indicated capacity. The temp. rise consideration of 55 deg. corresponds to this thought of overloading.

- Below is the table giving guidelines to decide the type of total system. Details are self-explanatory, and there are many more combinations for the designers to consider. We have considered a 1 MW Capacity plant as sample.

Table No.2

	Method of calculation of Number -of strings/ SCB and modules per string.				
S.No	ITEM	UNIT	FORMULA	VALUE	S.No.
1	Plant Capacity IN DC	MWp in watts		1000000	1
2	Nominal Power IN Watts	Wp		300	2
3	No of Modules	Nos	(1/2)	7352.941	3
4	Modules/String	Nos	Depends upon module Voc	22	4
5	No of Strings	Nos	(3/4)	1225.49	5
6	Strings per SCB	Nos	Decide	216	6
7	Total Number of SCB	Nos	(5/6)	5.673566	7
8	Module Specification VOLTAGE	Open Circuit Voltage	As per Spec.	159	8
9	Module Specification CURRENT	Current (Imp)	As per Spec.	1.3	9
10	Number of Inverters	Nos	As per Design.		10
11	One String of 22 Modules will contribute	Voltage	(4x8)	954	11
12	One String of 22 Modules will contribute	Current-Series Connection.	Same as 9	1.3	12
13	Current per SCB	Amps	(6x9)	280.8	13
14	Total SCB Current	Amps	(7x13)	1593.137	14
15	Total SCB Output or Inverter Input Voltage	Voltage-Parallel connection	Same as11	954	15
16	DC Voltage input per Inverter	Voltage-Parallel connection	Same as11	954	16
17	DC Current input per Inverter	Amps	(14/10)	1593.137	17

11.4 Civil Building Designing-Inverter Room

Apart from foundations of various equipments, the inverter room size and its air flow facility is one major aspect. Inverters are high heat producing equipments because of Diodes. If inverter room is not designed with air flow study, it will result in heating of the inverter internals, thereby reducing the efficiency of inverter, reducing power output and increase in losses. The inverter efficiency is directly proportional to the temperature inside the inverter. Hence inverter room air flow should be designed so that all hot air is removed continuously.

The heat produced in them touches 100 degrees Celsius in heat-sinks. This heat is sent out by means of their in-built exhaust fans. But still the rooms remain hot, and to cool them, free outside air should

be allowed inside the room continuously. Specially In the summer, high temperatures build up inside the rooms, which is the main cause of the burning of the Inverters.
Providing Air-Conditioning for inverters is not a practical proposition as the Air conditioner capacity will be very high, may be in the range of several tons.

Till now many inverters have burnt out in solar plants, only because of improper air displacement from the inverter rooms. Rate of air change for cooling air from the inverter room should be very high, as the inverter itself is a heat producing electrical equipment. By increasing the number of air changes to 120 per hour, hot air is exhausted (from the building as a whole and inverter in particular), and outside air is sucked in through open doors and windows. As a result the temperature inside the building is held close to the outside temperature. No matter outside temperature is 50° C. Our consideration is heat displacement. The rate of air changes per hour for cooling depends on the building construction and use:

Table No. 4.

Recommended air changes for cooling	
Conditions	**Air changes per hour**
Very warm workplace, offices, shops, etc.	60
Hot workplace, factory, workshop, warehouse, garage	80
Very hot workplace, bakeries, kitchens, laundries	120

The effectiveness of this type of cooling is dependent on the size and position of open loading bays, doors, windows, etc. If, inverters are throwing outside, through its ducts, x-meter cube of hot air per hour from their internals, then the window opening area can be calculated for an inlet air quantity equal to x-meter cube at a velocity of 1to 2meters per second.

11.4.2 SHADOW EFFECT-sun availability window with seasonal tilt-;

Next is the study of the shadow effect of any structure in the area like wall, poles tower, four pole, building etc. A day to day sun available window is to be calculated. The module rows are
to be placed at such distance from each other so that we get maximum sun window in all seasons.
Table below will help in deciding the max. Generation sun window in different months, date wise-;.

	Zero Shadow timings-Window				
	date	Tilt angle with horizontal in Deg.	No shadow time morning	No shadow time evening	No. Of Hours of Generation per day. Window
1	3-Mar	43	7:00	17:40	10:40
2	4-Mar	26	7:00	17:40	10:40
3	4-Apr	26	6:30	17:50	11:20
4	5-May	13	6:30	18:00	11:30

5	6-May	13	6:10	18:10	12:00
6	6-Jun	5	6:00	18:10	12:10
7	22-Jun	5	6:00	18:30	12:30
8	8-Aug	5	6:10	18:20	12:10
9	9-Aug	13	6:20	18:10	11:50
10	9-Sep	13	6:20	17:50	11:30
11	10-Sep	26	6:30	17:50	11:20
12	10-Oct	26	6:40	17:00	10:20
13	11-Oct	43	6:40	17:10	10:30
14	22-Dec to 3-Mar.	43	8:40	15:10	6:30

11.4.3 Sun Position Chart

This is the sun position chart for Kolayat area in Bikaner – Rajasthan-India. For designing the Modul layout we have to consider the shadow length of the modules mounted on the structures to decide the gap between rows of modules.

We have to select a window for maximum generation or peak generation, and accordingly decide the gap between rows of modules. You have to see that a compromise is made between maximum generation window and land cost, as more gap will consume more land. This chart shows the sun's position and shadow cast for 21-06-2013 for our plant location, i.e. Kolayat in Bikaner District, Rajasthan, India.

EXAMPLE

FOR SUMMER WE HAVE KEPT A PEAK GENERATION WINDOW OF 9 HOURS
FROM 8.30 AM TO 5 PM. PLEASE CHECK THAT THERE IS NO SHADOW EFFECT AND FAULTS DURING THIS PERIOD WHERE WE GET 95% GENERATION FOR THE DAY.

<u>Location</u> 27.830° N(Lat) (<u>Date:</u> 21-06-2013 <u>Time zone:</u> (GMT +5.5)
72.950° E(Long)

Table No. 7 Sun position table:

Local Time (GMT +5.5)	**Azimuth** (deg. from N)	**Altitude** (deg.)	**Shadow length** (multiplier)
05:45	62.946	RISE	-
06:00	64.672	2.451	23.363
06:30	67.920	8.523	6.673
07:00	70.945	14.732	3.803
07:30	73.801	21.052	2.598
08:00	76.540	27.462	1.924
08:30	79.217	33.945	1.486
09:00	81.894	40.486	1.171
09:30	84.654	47.071	0.930
10:00	87.619	53.687	0.735
10:30	91.000	60.317	0.570
11:00	95.233	66.938	0.426

11:30	101.408	73.501	0.296
12:00	113.273	79.847	0.179
12:30	150.642	85.004	0.087
13:00	225.909	83.824	0.108
13:30	251.252	78.060	0.211
14:00	260.697	71.618	0.332
14:30	266.119	65.031	0.466
15:00	270.037	58.404	0.615
15:30	273.268	51.776	0.788
16:00	276.157	45.168	0.994
16:30	278.883	38.594	1.253
17:00	281.553	32.068	1.596
17:30	284.242	25.605	2.087
18:00	287.009	19.219	2.869
18:30	289.908	12.928	4.356
19:00	292.992	6.756	8.441
19:30	296.316	0.729	78.590
19:37	297.132	SET	-

This page shows the sun's position and shadow cast for 22-12-2012 for plant location, Kolayat

FOR WINTER WE HAVE KEPT A PEAK GENERATION WINDOW OF 5 HOURS FROM 10.30 AM TO 3 PM.
PLEASE CHECK THAT THERE IS NO SHADOW EFFECT AND FAULTS
DURING THIS PERIOD WHERE WE GET 95% GENERATION FOR THE DAY.
Location: 27.820° N (Lat) 72.950° E(Long) Date: 21-12-2013 Time Zone: (GMT +5.5)

Table No. 8

Local Tim (GMT +5.5	**Azimuth** (deg. from N)	**Altitude** (deg.)	**Shadow length** (multiplier)
07:28	116.479	RISE	-
08:00	120.373	5.770	9.896
08:30	124.413	11.370	4.973
09:00	128.908	16.690	3.335
09:30	133.939	21.664	2.518
10:00	139.589	26.208	2.032
10:30	145.928	30.224	1.716
11:00	152.992	33.597	1.505
11:30	160.755	36.205	1.366
12:00	169.102	37.932	1.283
12:30	177.817	38.689	1.249
13:00	186.609	38.433	1.260
13:30	195.171	37.178	1.319

14:00	203.244	34.993	1.429
14:30	210.663	31.984	1.601
15:00	217.360	28.271	1.859
15:30	223.344	23.975	2.249
16:00	228.673	19.203	2.871
16:30	233.423	14.045	3.997
17:00	237.678	8.577	6.631
17:30	241.518	2.857	20.036
17:48	243.654	SET	-

11.4.4 The Plant Earthing and Lightening Protection Design.

PART A-Plant Earthing-

Please note that overall plant earthing for Direct Current area is not required.
This is explained here below.
To design the plant earthing system, the first step is the study of soil and site conditions.
The earth resistivity is the main factor-higher the earth resistivity-stronger will be the earth mesh.
The value of ER below 50 is a good value. The site conditions like weather-wet or dry, temperatures in summer, presence of moisture in earth, and at what depth, are main factors to be considered.
If the area falls under ISOCHRONOUS zone of the earth, where the % of lightning strikes are high, the earth mesh has to be very strong. Earthing resistance value has to be less than one ohm. As per the IEA it is a statutory requirement.

AIM –

- to kill the touch voltage in all metallic structures above the ground in the total plant area.
- to provide, less than one ohm earth resistance path, for every voltage surge and leakage current.
- to kill the direct lightning strike voltage, very fast.
- to kill the secondary internal switching and voltage surges in minimum time.

v.) Save cost of earthing mesh by using the piling as earth conductor.

The resistance of individual pile to earth is about 8 ohm. The reason is the cement concrete layer between the pile and earth. But if we connect all the piles together, through the super structure mounted on it, with GI strip, there will be a great number of parallel paths to earth. We know that -;
1/Req= 1/R1 X 1/R2 X 1/R3 and so on counting all legs of the support structures.
This will reduce the value of earthing resistance to less than 1-ohm. This value can be further reduced by strategically placing the earth electrodes of simple Perforated GI Pipe. To achieve this, the following steps a needed-;
Every structure is connected to its adjacent structure by welded GI strip of size 25X3 mm.
This means, the complete one row of structures of the total plant, should be inter-connected with the help of 25X3 GI strips. Therefore for this DC Area, we do not need any earth mesh separately. Same is true for the full plant structures.

Earthing Pits-;

The earth pit design consists of GI Perforated Pipe of length 3.00 meters. This pipe will be laid vertically in the ground by first digging the pit of depth 1.5 meters, ramming the pipe in this pit to one meter, then fill the pit with ordinary salt and coal powder for 300 mm and then fill the pit with loose sand.

Every SCB will have its own independent earth pit, if required by the SCADA engineer's special demand. There will be earth pits for four pole, inverters, transformers and LAs.

PART- B Lightening Protection.

The probability of indirect lightening effects occurring is significantly higher than that of a direct lightning strike. This is because every lightning strike within a one-kilometer radius can generate current flow in the modules, module cables and in the main DC cable by means of induction. Conductive and capacitive coupling are also possible and can equally cause overvoltage.

The lightning strike always travel to the highest point in the area of it's strike. Therefore the highest points-such as high voltage towers, poles, buildings and even the module supporting structure-is used by us to provide a protection from lightening.

Every module mounting structure leg also, is working as protection from lightening related damages. We cannot stop the lightning strike by any protection, but we can reduce the quantum of damag to our electrical equipments by sending the direct lightening surge, as well as the lightening induced surge in the electrical network, to ground within 3 milliseconds. This is achieved by providing LA on each four poles and SPD in each SCB. This fast killing of the voltage surge will save the plant and the persons working on it, from major damage.

We should provide simple 3 in. GI pipe on the roof of every building-control room, inverter room, placed vertically of a height 5 meters, which is connected to the ground earth point by a 25X3 GI flat. Please ensure that the earthing resistance of every metal part above ground in the whole park gives an earthing resistance of less than 1-ohm.in the entire AC Area of the plant

PLANT EARTHING AND THE LIGHTENING PROTECTION

The earth mat and the lightening protection system to be designed in a combined manner so that this protects our plant from all High Voltage Surges.

The HV surges in the system can develop because of two things-

i) The Switching (ON/OFF) of generators/feeders and grid side system operations.
ii) Lightening strike directly or in the vicinity.

Therefore the designed system should be capable to divert these surges to mother earth in micro-se ɔ save the plant from resulting overvoltage.

We already have LA and SURGE DIVERTERS embedded in our SCB, INVERTERS, HT Panels TRANSFORMERS and DP. All these require a zero resistance electrical path to earth to kill the HV Surge.

A. EARTH MAT-This path is given by the earth mat which is laid in the entire plant.
B. Every equipment body is connected to this earth mat (some call it earth mesh).

1. LIGHTENING ARRESTERS- For -LIGHTENING STRIKE protection- we provide copper alloy make high height much above the plant equipments. But these too are connected to earth- either directly or the earth mat.

2. EARTH MAT DESIGN CALCULATIONS FOR AC AREA OF PLANT ONLY

The earthing conductor sizing is designed based on the guidelines given in IS -3043 & CBIP manual with a fault current of 25k Amp for 1 sec.

A. INPUT PARAMETERS

1. Symmetrical fault current in substation (I) = 25000 A

2. Duration of fault current for sizing of conductor (t) = 1 sec.
3. Average soil resistivity (ρ) = **38 Ω.m**
4. Conductor type = GI
5. Initial temperature for material constants (θ1) =20 °C
6. Final temperature of conductor (θf) =500 °C
7. Reciprocal of temperature coefficient resistivity at 0°C for the conductor (B) =202 °C
8. Volumetric heat capacity of conductor material (Qc) =0.0038 j/°C.$mm3$
9. Electrical resistivity of conductor material at 20 °C (ζ20) = 0.000138 Ω.mm

$$S = \frac{I \times \sqrt{t}}{K} \text{ sq.mm}$$

Where k is constant which depends on the temperature capabilities of the chosen material of earthing which is **GI in this case.** K is calculated as below-

$$K = \sqrt{\left(\frac{Qc\times(B+20)}{\zeta 20}\right) \ln\left(1+\frac{\theta f-\theta 1}{B+\theta 1}\right)} = \sqrt{\left(\frac{0.0038\times(202+20)}{0.000138}\right) \ln\left(1+\frac{500-20}{202+20}\right)} = 83.9$$

Therefore "S" will be:- $S = \frac{25000 \times \sqrt{1}}{83.9} = 298$ sq.mm

We have selected the size as 50X6=300 mm for main earth mat and 25X6mm for connections of equipments to the earth-mat

B. <u>CORROSION ALLOWANCE</u>

As per CBIP Manual on Substation publication no 223, corrosion allowance is recommended as per following rule-:

a. In case of conductor to be laid in soil having resistivity greater than 100 Ω.m : no allowance
b. In case of conductor to be laid in soil having resistivity from 25 to 100 Ω.m : 15% allowance
c. In case of conductor to be laid in soil having resistivity lower than 25 Ω.m : 30% allowance

At our site the soil resistivity = 38 Ω.m
Hence 15% allowance is provided. Hence with corrosion cross section area will be
= 1.15 × **S** = 1.15 × 298 = **342.7 sqmm**
Therefore our earth conductor size selection of 50X6=300 sqmm is not OK. It has to be 50X8=400 sqmm. For 25 years life of the project.

C. <u>DESIGN</u>

<u>We have considered the following inputs specific to our plant-:</u>

1. Length of conductor buried in plant A.C area (L) = 3087 m
2. Depth of conductor buried in plant A.C area (h) = 0.3 m
3. Area occupied by conductor buried in plant A.C area (A) is divided into the following different areas-:
   a) Control Room
   b) Inverter Room
   c) Transformer Yard
   d) DP Area

   But the solar modules are covering the entire land area, which in totality is to be protected as every leg of the MODULE MOUNTING STRUCTURE and also, the support of each of the SCB is already earthed. So no earth mat is to be provided in this DC area. But its own earth mat using the leg supports is part of the earthing system.

Similarly every transformer, every inverter and all rooftops of each building are connected to earth. Therefore earth-mat will cover entire plant area which comes to =192030 sq met.for one MW

Therefore Resistance of total earth-mat (Rg) =

$$= \rho \times \left[\frac{1}{L} + \frac{1}{\sqrt{20} \times \sqrt{A}} \times \left(1 + \frac{1}{1 + h \times \frac{\sqrt{20}}{\sqrt{A}}}\right) \right]$$

$$= 38 \times \left[\frac{1}{4491} + \frac{1}{\sqrt{20 \times 192030}} \times \left(1 + \frac{1}{1 + 0.3 \times \frac{\sqrt{20}}{\sqrt{192030}}}\right)\right]$$

= 0.04718 Ω

D. <u>EARTH ELECTRODE AND EARTH-PIT DESIGN</u>

<u>We have considered the following inputs-:</u>

1. Soil resistivity of DP area (ρ) = 38 Ω.m
2. No. of ground electrodes (N) = 15
3. Length of each electrode (L) = 300 cm
4. Diameter of each electrode (D) = 10 cm.

Therefore Resistance of a single pipe or rod electrode , Re is-;

Re $= 100 \times \rho \times \ln\left(\frac{4L}{D}\right) / (2 \times 3.14 \times L)$ (where "ln" is natural log)

$= 100 \times 38 \times \ln(4 \times 300 \div 10) / (2 \times 3.14 \times 300)$

= 9.6563 Ω

Since there are 15 earth-pits, and since the system comes under parallel connection, therefore the effective earth resistance will reduce to 1/15times

RE = $resistance\ of\ single\ rod\ electrode$/no of rod electrode

= 9.6563/15

=0.64375 Ω

E. <u>FINAL RESULTANT EARTH-GRID RESISTANCE</u>

Since we are connecting the total earth-mat to all earth-pit electrodes, therefore the resultant earthing resistance will be: 1/R = 1/Rg + 1/RE

= 1/0.04718 + 1/0.64375 =0.044 Ω

Therefore Req. = 0.044 ohms

<u>Result:</u>

Since the Total earth grid resistance value is less than 1 ohm, hence the selected Earth Electrodes, Earth Mat Conductors, Protected Area and Quantity of Earth Pits are sufficient for application. There is margin of reducing the cross-section of the earth conductor along-with its length and also number of earth pits.

3. LIGHTENING ARRESTERS

i. If the plant area falls in the isochronous zone where the number of lightening strokes is very high, the entire plant area is to be covered with high mast arresters to protect the plant from direct lightening strokes. Each arrester covers a radius of chosen length 40/60/80/95 etc. meters hence the quantity is to be decided accordingly to cover total plant area.

ii. These LA will be connected to the earth mat or are earthed directly through the earth pit. this makes the total system compact and inter-related.

iii. If the area is in normal zone then there is no need to put the high mast LAs. The steel suppo structures act as the lightening protectors as their number is too large and the stroke current divides in these multiple contacts.

11.4.5 String Combiner Box

An actual picture of the internals of the String Combiner Box

- The modules are connected in series to form a string.
- The string voltage is equal to the multiple of Voc x no. of modules in a string. It should be less than 1000 Volt.
- The string current is equal to the individual module current because of series connection of module
- Strings of equal length are then connected in parallel to make up the PV generator, where the outpu power of the strings is cumulative. Here the limitation is Voltage less than 1000 Volt and current less than 1000 Amp.
- Multiple string cables from the PV generator are consolidated using necessary fuses and combining bus bars in the String Combiner Box (SCB).
- The SCB/DCJB is located close to the modules and connects several strings in parallel, meaning that only one positive and one negative cable with higher cable cross sections must be laid from each junction box to the downstream inverter.
- This cable from each SCB carries the solar power of PV generator, to the Inverter. It can also perform additional safety-related functions, such as those of string fuses or overvoltage surge protectors.

- If thin-film modules are used which are not reverse current proof, blocking diodes must also be employed. If not, then low rating fuse(equal to the Isc) should be used, which will work as blocking diode.
- In addition, there are certain components which may be provided within the SCB. For example, the main DC switch could be a part of the SCB. Also surge protection device (SPD) is also embedded in the SCB.

Finally therefore, to sum up, the Solar Power generator chain can be explained as below:-

-Mini solar cells are embedded in the module.

-These mini solar cells are connected in series and are protected by a container called a module.

-Modules are connected in series to a make a string, of course subjected to voltage and current limitations.

-Several strings (normally 22 or 24) are connected in parallel inside a String Combiner Box (SCB)

-Number of SCB (depending upon the system design) collect power from the strings and feed it to the Inverter. Normally for 1 MW generator the SCB are 7 to 8 nos.

-Finally the entire solar power which is generated by mini solar cells- is collected via the series and parallel connections and fed to the inverter.

PRACTICAL CONSTRUCTION PLAN for MW CAPACITY SOLAR PV PLANT

Detailed below is a PRACTICAL construction plan for the construction of a MW capacity solar PV power generating project. This plan is prepared after taking feedback from site engineers who have carried out the works of plant erection testing and commissioning. While preparing this plan we have learnt from their successes and their failures. In this report we have also shared our good and bad experiences with-

- the equipment suppliers,
- their commissioning engineers
- their supplied equipments
- our own designers
- our own procurement engineers
- our own site engineers
- Contractors and their workers and many more.

After reading these guidelines, we hope that efforts will be made by all concerned, to improve the working in all fronts and areas detailed therein.

1.0 CONSTRUCTION PLAN Part-1

The overall work of construction of a solar power generating plant starts from the date of its inception. The ETC of the electrical equipments is a complicated affair, mainly because many agencies are involved in the total work. Close coordination is required between the various agencies viz.

- Designers,
- Procurement Team,
- Erection team
- Contractors and the workers

And above all the Equipment suppliers

Therefore, it is necessary to prepare a PRACTICAL operating plan and then to follow this plan strictly.

We start from the beginning, that is, when it is finally decided that the plant will be of capacity

- ✓ Say M-MW,
- ✓ the planned target of generation is G-KWH
- ✓ The anticipated cost of the plant is R-Rs/MW
- ✓ The commissioning target of the plant is D-Days

Once the above is finalized and a Go-ahead is given, the next step is the fixing of timelines for each activity by the top level. The time lines will contain both-the commercial aspect as well as the technical aspect. The target date of commissioning of the plant is indicated in the timelines. The format is as below-:

2.0 PERT CHART---- Because of cost considerations it is essential that the solar power plant is constructed in the minimum time. Below is a pert chart format for construction of a solar power project in India. All activities have been filled in this table in Col-1. All concerned must sit in a PRESTART meeting and decide the dates, days and time taken planning for each activity-;

Pert Chart for the Project Planning, designing, procurement, erection, testing and commissioning of the plant. The target period for completion is 120 to 150 days for the complete works.

Name and capacity of the Project- --

Sl No.	Task Name	Start Date	Finish Date	Duration Days	Comments
1	**Project schedule for 5.00 MW solar PV project**				
2	Receipt of order with financial closure				
3	Takeover of the land				
5	Access road to site (Customer's scope)				
7	**Statutory approvals(Customer's scope)**				
8	CEIG(Design)				
9	PCB/ Environmental Clearance./Factory Inspectorate				
10	**Design & Engineering-Basic and detailed Design**				
11	Soil investigation				
12	Contour survey				
15	PV Design				
16	Electrical design				
17	Layout				
19	PV Design				
20	Electrical design				
21	Layout				

23	**Release of Final drawings & documents to procurement wing and site staff**				
24	Purchase specifications				
25	Civil				
26	Electrical				
27	Instrumentation				
28	MNRE Approval – For duty exemption				
29	**Procurement and Material Management Activity- Tendering, Negotiations, Ordering, Inspection of ordered Material**				
31	Modules				
32	Module mounting structure				
33	Inverters				
34	String combiner boxes				
35	Control & Monitoring equipments				
36	HT panels				
37	Transformers				
38	Double pole structure				
40	Power cables & accessories				
41	Control cables & accessories				
42	Earthing materials & Lightning protectic system				
43	Lightings & accessories				
44	Electrical contractor				
45	Civil contractor				
46	Mechanical contractor				
47	**Receipt of material at site and the checking that correct material is received**				
48	Modules				
49	Module mounting structure				
50	Inverters				
51	DC combiner boxes				
52	HT panels				
53	Transformers				
54	Double pole structure				
55	OH Line Material				
56	Power cables & accessories				
57	Control cables & accessories				
58	Earthing materials & Lightning protection system				
59	Lightings & accessories				

60	Control & Monitoring equipments				
61	SCADA				
62	**Civil works, Installation & Commissioning**				
63	**Selection of contractors and placement of orders on them**				
64	Setting up of site office				
65	Availability of power for construction(Customer's Scope)				
66	Mobilization of civil team				
67	Bore-well works				
68	Internal roads & Drains				
69	Boundary wall				
70	Excavation & Foundation for module mounting structures				
71	Cable trenches				
72	**Installation**				
73	Installation of module mounting structures				
74	Installation of modules & interconnectic of modules				
75	**Control room**				
76	Excavation				
77	Foundation				
78	Column, beams & slab work, brick work finishing work				
79	**Installation**				
80	Electrical panels				
81	Inverters				
82	DC combiner box				
83	Battery bank with charger				
84	**Power Cable laying & termination**				
85	Wiring of modules till DC combiner				
86	Laying of cables from DC combiner t inverter room for modules				
87	Inverter room wiring				
88	Wiring from inverter to Transformers				
89	Wiring from Transformer to HT Panel				
90	**Lighting, Earthing& Lightening protection**				
91	Lightings & accessories				

92	Earthing				
93	Installation of lightening protection system				
94	**Instrumentation and Metering**				
95	Installation of SCADA & control equipments				
96	Control cable laying & termination				
97	Providing Energy meters as per DISCOM				
98	**Transformer & Double pole structure**				
99	Excavation & Foundation for transformers				
100	Installation of transformers				
101	Excavation & foundation for double po structure				
102	Installation of double pole structure				
103	**READYNESS OF THE PLANT FOR COMMISSIONING-120th DAY FROM 00 DATE**				
104	**Testing**				
105	Modules				
106	DC combiner boxes				
107	HT Electrical panels				
108	Inverter				
109	Battery				
110	Transformers				
111	SCADA				
112	Metering as per DISCOM norm				
	Commissioning				
113	**Availability of transmission lines up To substation.**		**Separate Activity depending upon the agency who will do this work**		
114	Back-charging of Transformers				
115	Extending power supply to Inverters				
116	**Commissioning**				
117	**Start of Power evacuation**				
118	**Hand Over of plant to customer**				
119	**ANTICIPATED DELAY BECAUSE OF ANY REASON=20 days**				

3.0 Start of Works.

Once the above PERT is discussed in its minutest detail, the actions by the involved agencies will start right away. The agency wise work details are as below.

3.1 Design-Engineering

Once these features are decided, the design team has to take the first step of designing the total project concept and preparing the following drawings -

- Land layout indicating approach roads etc.

- Topographical and contour survey report.
- Soil investigation and earth resistivity report
- The GA layout of the plant, in which, each and every equipment is indicated, along with their short specifications.
- The electrical single line diagram is prepared.
- Control room and inverter room design with drawings.
- Cable routing layout.
- Auxiliary power supply design and drawing.
- Power pooling design and drawing.
- Power evacuation plan/design/drawing.
- Rest of the drawings will follow, as and when the scheme is finalized.
- Once the preliminary scheme is checked /approved/ by the Director, all the drawings and design data will be freeze and released for procurement, for site works and for commercial aspects.

3.2 Bill of Material -; After the finalization of the scheme, the designer will prepare the detailed bill of material (BOM) for each and every item required for complete project. He will also prepare the specifications of the items to be procured. This BOM will be in a format, as detailed below.**PRACTICAL Cost of materials purchased in recent past can also be indicated in this table.**

Preliminary BOM				
Sl No.	**Item**	**General Technical Specifications the values given below are indicative only and correct specifications based on th equipments selected are to be filled in this column.**	**Unit**	
A	**Solar Field-DC area equipments**			
1	Module Mountir Structure	Suitably designed to bear the load of modules with a layer of one module in a vertical column. Hot Dipped Galvanized Steel Structure	Kg	
2	Solar Modules	Crystalline-with specific mention that the modules are free from PID. 295Wp.	Nos.	
3	4 sqmm DC cable	1CX4 sqmm, stranded copper, un armored, suitable for outdoor mounting and exposed to hot direct sun light with anticipated temperature rise to 80 deg. cent.	m	
4	String Combiner Box	Outdoor type, IP 54,for 24 inputs, with fuses on both +ve and -ve, for 1000 volt DC, having Surge protection,	Nos.	
5	MC4 Connectors	Connector Pair (Male & Female)	Nos.	
6	DC Power cable-1Cx150sqmm	1Cx150 sqmm, stranded aluminum, armored,	m	

7	Inverters	630 kva, self-cooling, no forced air required, with input DC current as1000 amp and DC input Voltage as 1000, giving AC output at 0.350 k v	Nos.	
B	**AC AREA Equipments**			
8	HT cable joinir kit	185 sq mm HT cable joining kit straight through and end	KIT	
9	Angle Iron	50x50x6 mm angle	KG	
10	ISMC-	ISMC-125x50	kg	
11	LT cable	2cx2.5 sqmm LT cable	m	
12	power cable-	Aluminum, stranded,1.1 kv, armored, XLPE, 3CX300	Met.	
13	Power Transformer.	.350-.350/33kv,oil filled, ONAN, off-circuit tap changer,Yn-YnD-11,790-790/1580kva, three winding copper wound transformer, complete with Bucholz, OTI, WTI, PRV, MOG, completely wired Marshalling Box	Nos.	
14	HT power cable-	33 kv, Aluminum, armored 3CX185	M	
15	33kv VCB	Outdoor type, 33 kv VCB with mounting structure and power pack arrangement. Complete with C&R Panel containing all protection	Nos.	
16	33kvCurrent Transformer	50/5-5 amp.- 4nos. & 150/5-5Amp- 1no. Of two core, oil filled, copper wound, 5p20 class, outdoor mounting with mounting structure	Nos	
17	33kvPotential Transformer	33kv/110-110volt,oil filled,5p20,outdoor type, copper wound	Nos.	
18	33kvLightening arrestor	33 kv, station class	Set	
19	33 kv Four Po Structure	As per design	Nos.	
20	Plant Earthing Electrode and pits	25x3 mm GI flat, 50x8 mm GI flat Earth electrodes, 3meter deep with perforated pipe and chemicals.	Km/mw	
21	33/.415 Aux.Transformer	25 kva,33/.415 volt, copper wound oil filled, off circuit tap changer, with cable termination on the LV side	Nos.	
22	415 volt ACDB	415 volt, one input and5 outputs for 3-phase and 5 outputs for 1-phase.outdoor pole	Nos.	

		mounted, aluminum bus bar		
23	Ferrules, Gland Lugs,	All relevant sizes	L.S	
24	HDPE pipes different sizes	As required	M	
C	**Other Equipments**			
25	SCADA system		Set	
26	Plant Lighting	20-Watt,solar + ac lighting	Nos.	
27	Fire Extinguisher 2.5 KG		Nos	
28	Rubber Mat		kg	
D	**Civil works**			
29	Land leveling		M	
30	Boundary Wall/Fencing		Km	
31	Pathways and Roads		km	
32	MMS fixing		mw	
33	Transformer Foundations		nos	
34	Other equipment foundations		nos	
35	Inverter Rooms		Sqft	
36	Main Control Roor		Sqft	
37	Ac side fencing		nos	

- The BOM should contain basic specification of the equipment. Separate sheet has to be attached for detailed specifications.
- This BOM and Specs will be sent to the procurement team for arranging purchase.

3.3 Procurement-Material Management

- Once the BOM is finalized, the procurement team will float the tender enquiry to a minimum of three suppliers.
- This team should be an experienced team, which should be equipped with the list of manufacturers and suppliers of the equipment, along with their telephone nos. and e-mail address.
- This team will share all technical inputs with the designer and the technical team.

- Suitability of the equipment on offer is to be got confirmed from the concerned technical experts, before placing the orders for supply.
- The commercial aspects are also to be settled by the procurement team in consultation with the top management.
- The timelines about the availability of the equipments is to be prepared and sent to the field engineers so that they can plan the site works according to the receipt of material.
- As far as possible, the best available equipments are to be procured. The grading of the equipments is well-known, and decision for selecting the equipments needs to be taken after discussions with the designer and technical team.

3.4 Placing of orders for the supply of material

It is observed everywhere that proper purchase order is not placed on the suppliers. Once the supplier gets the order, he does not bother about maintaining the quality of the material, and in case of dispute at a later stage, he takes advantage of the weaknesses left in the PO. It is therefore necessary to place the order, which should contain the following details positively-;

Free of cost or cost included provisions for- “Assistance during Erection, Testing and commissioning of the supplied equipment.”

This aspect is never mentioned in the PO, and the supplier does not provide this facility. Result-the equipment remains partly commissioned or un-commissioned for long periods. This is especially for the VCB and related power pack, CR Panel, SCADA and Inverters.

In electrical circuits, the termination details of all the cables in the equipments is always not discussed/examined, resulting in the problem at the time of installation. It is absolutely necessary that the equipment supplier should know beforehand as to what size and numbers of cables will be connected to the equipment terminals.

- Warranty
- Payment terms-efforts should be made to detain 10% of the contract value as security.
- Delivery timelines and LD clause.
- Order should be examined by the accounts wing, to check that all conditions are met in the draft of the order.
- Service after Sales details should be mentioned in detail in the order.
- The copy of the PO should invariably be sent to-the designer, accounts wing, storekeeper, site engineer.
- The PO should contain the provisions of Transit Insurance, which will be done by the supplier at our cost.

3.5. Tests and inspection of material at the supplier’s works.

The test reports and testing of the material is often avoided, because this consumes a lot of time. This is not correct especially in case of main equipments.

- Test certificates should be examined in reference to the ISS or international PRACTICALs.
- Type tests cannot be got done but we should insist on the routine tests-both at the works and also at site.

3.6. Receipt of material at site.

- The storekeeper at site should examine, that he has received the correct material at site as per the PO.
- Site erection engineer should also check the material for its correctness.
- The material is received in good condition.
- It is stored properly as per the guidelines of the suppliers.
- Final payment to the supplier should be released only when the storekeeper gives clearance after examining the material received.

3.7. Works at the Site

The site works can be divided into the following sub-sections

- Site Preparations.
- Selection of the contractors.
- Start of the civil works.
- Start of the equipment installation works
- Cabling
- Termination/Interconnections of the electrical circuitry for - power as well as control and protections.
- Testing of the installed equipments.
- Commissioning of equipments in steps.

4.0 Site Preparation and Selection of the contractors

a. Once the project is cleared for implementation, site preparation is to start.

- In this, first step is of posting proper-qualified and experienced site engineers of both trades electrical as well as civil. Work distribution of engineers should be made so that the each engineer is as assigned with the particular work.
- The concerned engineer will collect all the drawings and related literature of all equipments, in both soft copy and hard copy. He will study this so that proper erection of the equipment is carried out.
- Experienced and financially sound contractors are then to be selected. Only that contractor, who has required tools and plants, engineers and diploma holders in his employment and has sound technical and financial capabilities, should be employed. This selection criterion is the most important point for a plant to be constructed to give the desired output satisfactorily. Any cost saving, while selecting the contractors at any stage, causes heavy loss during O & M of the plant. Also shabby and dirty plant will be constructed.

b. Start of the civil works

- The civil works will start with construction of approach role and internal pathways. Ground leveling, drainage system, boundary wall ware house and security gates are to be arranged through good quality civil contractors. It will be better if independent contractors for each work are employed so that work quality is maintained.

c. Equipment installation works

- Once the foundation of main equipments is made, the engineer should plan for the installation of the main equipments. For this the engineer should properly read the guidelines for installation of equipments as given in their documents. Wherever required the suppliers engineers should be called.
- The inverters, transformers, SCB and the modules are delicate equipments and may damage if improperly handled. And therefore these should be erected with full care by employing experienced labor so that no damage is done due to careless handling.

- These equipments are prone to damage by rain water and heavy dust therefore proper covering is absolutely necessary during the course of erection.
- Arrange site routine testing of the equipments –transformers, inverters, VCB with CR Panel, SCB etc.
- Check the Insulation Resistance of the cables, equipments and record the values

d. Cabling

There are four types of cables used in the plants viz.

HT power cables,

LT power cables

DC power cable

SCADA communication cables

Most of the cables are laid under the ground. Therefore to avoid damages to the cable, the designer has indicated the cable routes for each type of cable. Cable trenches should be dug up to the depth of minimum 1 meter.

The armored cables can be buried in the ground directly. The unarmored cables should be laid in PVC or HDPE pipes for protection from damages and also for rodent protection.

The cable route should be marked in the plant in two ways:-

- ✓ by white, or colored powder, over the ground.
- ✓ by cable route markers of stone/metal placed on the route at a distance of 30 feet each.

5.0. Erection, Testing and Commissioning

Finally, when all the equipment are installed, cables are laid, terminations are done and plant is ready for commissioning, then site engineers will sit together and choke out a plan for the commissioning activities. For this, the following steps are necessary-

- Check that the statutory approvals are available for the plant, line and all related equipments.
- Earth resistance of the earth mesh and also of the body of the each equipment is measured and recorded.
- The equipment supplier's engineers are available for helping in the commissioning.
- Site test reports of each of the equipment is checked and found OK.
- Give go ahead for plant commissioning.

The second part of this guideline follows here below which contains the equipment wise technical guidelines and checks required to be carried out by field engineers.

6.00 Detailed below are the item wise guidelines for checks for the construction activity-;

- ALL ELECTRICAL ENGINEERS AT SITE SHOULD INVOLVE FULLY WITH THE SUPPLIER'S COMMISIONING ENGINEERS LEARN THE TECHNICAL INTRICACIES OF THE EQUIPMENTS SUPPLIED BY THEM. THE TECHNICAL SPECIFICATIONS SUPPLIED BY THEM TOO SHOULD BE STUDIED AND DISCUSSED AT SITE SO THAT THEY LEARN TO OPERATE AND MAINTAIN THE EQUIPMENTS IN QUESTION.
- ALL CIVIL ENGINEERS SHOULD FOLLOW THE BASICS OF THE CIVIL ENGINEERING-AS THE WORST ENEMY OF A SOLAR PV PLANT IS THE WIND AND STORMS- EVERY BUILDING AND STRUCTURE SHOULD BE DESIGNED AND CONSTRUCTED TO WITHSTAND THIS FORCE FOR A WIND VELOCITY OF 150 KMPH.

The supplier and manufacturer gives complete guidelines for the site handling of their equipments which should be available to site erection engineer. He should strictly follow these guidelines. In case of any doubt he should contact the suppliers engineers. For main equipments, it is desirable that the supplier's supervisors are called to look after the erection work.

Sl. No	Item	Checks and Procedures-Please read "check that" before each item here below-:
6.1	Material Receipt in site store	❖ **Store keeper must have a copy of the Purchase Order placed on the supplier. He should check the Manufacturer's name & material details supplied by him against the PO. The site incharge of construction should have the copy of the specifications of the equipment received, and he should confirm that the item received is as per the design and order.**
		❖ Re-Check that a correct material has been received.
		❖ Ensure Size, length, Specifications & Quantity of each and every item, received in the store, including Channels, Angles, Brackets, Transformers, inverters, Modules, VCB, CT,PT and confirm that the items are as per the order placed and confirms to the BOQ & Specifications
		❖ Ensure that there is no physical damage of the received materials & its accessories at the time of transportation, shifting & un-loading. If there is any damage, then the report with photograph should be sent to the HO immediately.
		❖ Identification tag shall be tied indicating material details and it will be entered in the stock register
6.2	MODULE MOUNTING STRUCTURE	❖ The piling work or foundation is similar/common in all types of MMS.
		❖ The site engineer has to personally check each and every pile starting from digging of the pile hole, placement of column, check of its depth and verticality, ratio and quality of cement concrete mixture, height, above ground of its top, alignment with the complete series of columns and any other aspect as per level and clearance considerations.
		❖ The **True North** should be determined and the EAST-WEST axis be drawn physically on the ground.
		❖ The channels, Angles, Brackets & other accessories are received as per the Specifications & BOQ.
		❖ before installation, channel & Angle supports must be straightened
		❖ The foundation of pile are as per the approved drawings/specifications
		❖ The height and distance between pile are as per the approved drawings/specifications
		❖ The foundation and piles are in one straight line as per the approved drawings/specification
		❖ The suitable size of base plates are fixed above the pile, **if it is two part structure.**
		❖ Suitable size of anchor fasteners are fixed for the base plates, **if it**

		is two part structure.
		❖ Height of vertical -channel supports are as per the drawing/specification
		❖ Continuous welding done for vertical -channel support to the base plates
		❖ Grinding, finishing & painting done for the welded portion of the channels and base plates
		❖ In case the column is itself grouted in the ground, then its verticality is to be checked with spirit level and plumb.
		❖ Suitable size and length of raft and perlins are provided as per the design of MMS.
		❖ That suitable size of nuts & bolts are fixed for the angle supports
		❖ The alignment of channels, cross arm angles and leg with water pipe, plumb, or sprit level has been done.
		❖ the final tightness done for the joint plate's Nut & Bolts
		❖ Foundation of plinths are as per the approved drawings/specifications
		❖ The finishing of concrete plinth's as per the drawings/specifications & per site condition.
6.3	**Modules**	❖ all parts of the modules, during transport and installation, are free from mechanical stress (i.e., from pressure & tension.
		❖ the final alignment of structure channels, cross arm angles and rails with water plumb level/sprit level has been done to prove its correctness.
		❖ only undamaged solar modules are installed
		❖ the module is placed in proper portrait, with positive on top and negative on the bottom-this means all modules should be placed similarly. Or otherwise for other configurations like landscape the terminals should be on same directions for all modules.
6.4	**String Combiner Box**	❖ the fixing of GI channels for the combiner boxes as per the Drawings/Specifications.
		❖ Clamps supports are fixed for the combiner boxes as per the drawings/Specification. Nut & bolts holes are provided at the base of the combiner boxes. the tightness of the clamp support nut, bolts of the combiner boxes the tightness of the clamp support nut, bolts of the combiner boxes.
		❖ 4 sqmm flexible wires must be connected from each solar module string to a combiner box as per the drawings/SLD
		❖ the tightness of weather proof glands of the combiner boxes
		❖ there is no straight through joints in the wires between the module junction box to combiner boxes. If it is required, then it should be proper open to weather electrical joint.
		❖ identification ferrules & lugs are provided for the wires at the termination ends
		❖ Polarity of the wires in the combiner boxes at the termination ends, select colour code for +ve and –ve cables.
		❖ the rating of the Fuse is as per the drawings/SLD/design

		❖ each series string of panels to be wired to its own +ve & -ve fuse as per the drawing/SLD
		❖ tightness of the wires at the termination ends. MC 4 connecters are used and properly crimped.
		❖ commissioning of the combiner boxes must be done by experienced engineers.
		❖ that the IP protection is followed so that there is no ingress of solids and water during worst weather.
6.5	**Inverter**	❖ Each panel location, as per the drawing, and layout for erection has been arranged. The room clearance of each panel & panel base frame should not be less than permissible and should be as per the design. The alignment of the panels with water tube/level guage/ plumb level should be done.
		❖ each & every equipment in panels are as per the approved drawings
		❖ that the cooling air flow is not restricted and there is ample room for the movement of O&M staff inside the IR. Provisions of changing/cleaning the air filters is provided in the windows/openings.
		❖ all panels are properly bolted in floors with suitable anchor fasteners
		❖ tightness of all panel joints, nuts and bolts, clamps and connecting terminals & termination point shall be checked with the help of torque wrench spanners
		❖ spring washers are provided for the nut & bolts at the termination ends
		❖ suitable supports and clamps are provided for the cable above or below the panels
		❖ all the panels are earthed at two ends.
		❖ unwanted holes are closed on top & bottom of the panels
		❖ carry out the insulation resistance test as per the manufacturer's guidelines and record the values.
6.6	**Transformer**	❖ the base foundation of the transformer is as per the approved drawing and technical specification. Rails & rollers are intact before shifting of the transformer. The position of the transformer is properly erected on the steel base inserted in the concrete plinth foundation. The stoppers supports are provided for the wheels to arrest the movement of the transformer
		❖ if there are more than one transformer in the plant, the transformers should always be separated from one another & also from walls for free circulation of cooling natural air
		❖ the body earthing & neutral earthings are done as per the approved earthing drawings & specifications
		❖ the appropriate rating of cables are connected as per the drawings. Suitable size & adequate number of supports are provided for the cables if required.
		❖

		❖ all the control wirings are done as per the control drawing
		❖ the cable entries to the transformer are provided with proper sealing for adequate protection against the ingress of moisture, insects and reptiles..
		❖ The IR value should be taken between HV to earth, LV to earth and also HV to LV. BDV of oil be checked and recorded.
6.7	**HT Switchgear**	❖ at the outset check that the panel requires what type of auxiliary supply-is it DC, is it AC or it has power pack backed up by AC 220 volt. supply, or is it 110 AC from PT.
		❖ the base structures are fabricated and installed as per the layout of switchgears and that position of the Switchgears are properly fixed on the steel base inserted in the plinth foundation and the switchgears are properly bolted in base angle supports and the switchgear foundation's plinths shall be minimum 300mm above finished ground level
		❖ that the cleaning of the switchgear before erection has been done properly and no foreign material is left inside any opening and chamber.
		❖ the switch gears are coupled as per the GA drawing
		❖ all the doors, other than cable chamber, shall be hinged & lockable type with neoprene gaskets at all the joints and the cable chamber shall be fixed with nut & bolts
		❖ the wire mesh guards to stop entry of reptiles & gaskets shall be provided for cable chamber
		❖ the tightness of all the bus bar joints, nuts, bolts, clamps, connecting terminals & termination point with the help of torque wrench/spanners as per manufacturer specifications.
		❖ all the control wirings are connected as per the control wiring drawing
		❖ all the electrical and mechanical operations of the breakers, meters & relays are ok.
		❖ all the empty holes are closed from top & bottom of the switchgear
		❖ the earthing connected at both end of the switchgears
		❖ each feeder shall be provided with a stainless steel name plate
		❖ the switchboard shall be comprising of incomer feeder, bus PT & outgoing feeders and are marked accordingly.
		❖ all mechanical indications of breaker truck, carriage positions, breaker ON/OFF trip conditions, spring charged/ discharged indications shall be freely visible from outside and easily approachable. Emergency trip push button shall be visible from outside.
		❖ Height of operating handles of control switches, push buttons, and reset rods of relays shall be as per operation norms.
		❖ the Circuit breaker with its control units, operating mechanism, isolating and interlocking gears, auxiliary switches, isolating contacts and wiring is fixed on a horizontally with-draw able, sheet steel truck/carriage on wheels. Therefore there should be ample space in

		front to draw out the breaker.
		❖ the mechanical interlocks shall be provided on each truck carriage to prevent mal-operation
		❖ the trench/duct shall be suitable for terminating cables or bus turnings as specified, sufficient space and support arrangements shall be provided for terminating specified no of power cables with bottom entry.
		❖ the tightness of nut & bolts at all the bus bar & termination joints in the switchgear panels with the help of torque wrench spanners as per manufacturer specification
		❖ the continuity of all the bus bars & earth continuity inductor provided in the switchgear is available.
		❖ the two earth terminals shall be provided on each switch board as per the approved earthling drawing
		❖ the vacuum cleaning done for the switchgears.
6.8	**HT Control and Relay Panels**	❖ Please check at the outset that the panel requires what type of auxiliary supply-is it DC, is it AC or it has power pack backed up by AC 220 volt. supply
		❖ the base structures are fabricated and installed as per the layout of switchgears and the panel base frames are not less than 75mm in height. Position of the control & Relay Panel are property arrested on the steel base inserted in the plinth foundation.
		❖ cleaning of the control & Relay Panel before erection using air blower.
		❖ all the doors other than cable chamber shall be hinged & lockable type with neoprene gaskets at all the joints and the cable chamber shall be fixed with nut & bolts
		❖ the additional wire mesh guards & gaskets shall be provided for cable chamber to stop any ingress. all the empty holes are closed with gasket from top & bottom of the control & Relay Panel.
		❖ all the control wirings are connected as per the control scheme.
		❖ all the electrical and mechanical operations of the breakers, meters & relays are OK
		❖
		❖ that the earthling connected at both ends of the panel is provided.
		❖ the control & Relay Panel shall be provided with a stainless steel name plate
		❖ that the insulation resistance test to be carried out for the control & Relay Panel and the readings to be recorded
		❖ the type for indoor or outdoor installation and its degree of protection IP be checked.
		❖ doors shall be of hinged and lockable type with neoprene gaskets at all joints and the cable hamper shall be fixed with nuts and bolts
		❖ All indicating Instruments shall conform to the requirements specified in design criteria.

		❖ The scheme of connection of all equipments at the substation should be as per the design and indicated on the MIMIC diagram. Also the circuit breaker and isolator control switches and semaphore indicators shall be incorporated in the Mimic diagrams.
		❖
6.9	**Outdoor Circuit Breakers**	❖ the civil foundation base of the. structures are constructed as per the approved drawing & plinth shall be 300mm. height above the ground level
		❖ all the GI structure materials shall be straight & free from bend or twists and all burrs and irregular edges will be ground smooth
		❖ the vertical alignment of CB is checked with water & plumb level Ensure the height of the bushing/support insulator of the CB should be as per Indian electricity act
		❖ the exposed live parts of the CB shall be placed high enough above ground to meet the IE act requirements.
		❖ON/OFF/TRIP indication lamps, along with a mechanical 'open/close' position indicator, should be visible to the operator standing on the ground.
		❖ the distance between the isolators shall be as per specifications/drawings
		❖ the base channels &other structural steel members such as operating pipes, phase coupling rods, operating mechanism boxes, bolts, nuts & pins etc. are fixed properly. The bolts, screws & pins shall be provided with lock washers, key or equivalent locking facilities.
		❖ the supporting structure of the CB shall be of hot-dip galvanized steel
		❖ bi-metallic connectors must be used for any connection between dissimilar materials
		❖ the structures, insulators, fixed & moving contacts are installed as per the approved drawing & as per the specifications, Indian and international practices.
		❖ the pressure & dielectric strength of the insulating gas shall be checked after installation, if required SF6 shall be filled up to the pressure as per supplier's manual
		❖ anti-corrosive grease shall be applied at bimetallic connections
		❖ all the joints & coupling of the structure parts are done with suitable bolt with a steel washer under the nut.
		❖ the earthing done from top to bottom of the structure and to the earthing pit or earth grid
		❖ the circuit breaker shall be suitable for remote control from the electrical control room & check the local control switch with locking arrangement shall be provided for each breaker for local operations
		❖ the facilities for remote indication of breaker 'open/close' position in the electrical room
		❖ the cabinet for operating mechanism and its accessories shall be of IP55 protection with pad locking facility
		❖ all the iron parts shall be hot dip galvanized

6.10	**Isolator with earth switch**	❖ for the erection of pole, check the excavation of pit length, depth, width & base padding are done as per the approved drawings. All the relevant support fittings fixtures & conductors shall be as per IS
		❖
		❖ position status with earth switch structure property erected on the steel base inserted In the concrete plinth foundation. The alignment of the Isolators with earth switch structures is properly aligned.
		❖ the inter phase clearance and mounting height shall be more than 1.5mtr or as per the IE Act requirement.
		❖ the earthing is connected to both the poles of the Isolators with earth switch
		❖ all the bolts are tightened with torque wrench spanners
		❖ the each isolator with earth switch structures shall be provided with a stainless steel name plate
		❖ bi-metallic connectors must be used for any connection
		❖ all the GI structure materials shall be straight & free from bender twists and all burrs and irregular edges will be ground smooth
		❖ the levelness & tightness of the insulating base of isolator with earth switch structures
		❖ the distance between the isolators shall be as per design.
		❖ the steel structures should be at designed height
		❖ the height of the bushing/support insulator should be at designed height from base channel
		❖ the clearance between the ground and the live conductor portion is as per design.
		❖ flat washers shall be circular & of suitable thickness and spring washers of suitable size
		❖ the speed of operation during opening or losing shall ensure minimum arching
		❖ each pole of the isolator shall be provided with two earthing base
		❖ the flexible tinned copper earth connectors shall be provided for the earthing system.
		❖ the disconnecting switches shall be provided with earthing switches on one side forming integral part
		❖ the ganging mechanism shall be provided with sufficient adjustment to allow for final alignment of the isolator blades
		❖ the manual operating handles shall be mounted on the base of the supporting structure. Guide bearings shall be provided, if necessary at a designed height.
		❖ the isolator and earthing switches shall be provided with padlocking facility to lock them in fully open or fully closed positions
		❖ the isolator and earthing switches shall be mechanically interlocked (castle key interlock) such that it will not be possible to close the earthing switch when the isolator is closed and vice versa.

		❖ the: civil (foundation base of the structures are constructed as per the approved drawing & plinth shall be 300mm height above the ground level
		❖ the GI structure materials shall be straight & free from bend or twists and all burrs and irregular edges will be ground smooth
		❖ the vertical alignment of CT and PTs-structures with water & plumb level
		❖ the level & tightness of the insulating base of CT and PTs
		❖ the height of the bushing/support insulator should be as per specifications
		❖ the clearance between the ground and the live conductor should be as per design.
		❖ all the joints & coupling of the structure parts are done with suitable steel washer under the nut.
	Potential and Current Transformers	❖ earthing done from top to bottom of the structure and to the earthing pit or earth grid
		❖ wiring of the CT andPTs secondary box and the fuses provided shall be checked after erection.
		❖ the bolts, screws & pins shall be provided with lock washers, key or equivalent looking facilities
		❖ bimetallic connectors must be used for any connection between dissimilar materials
		❖ the polarity of primary and secondary terminals of CT should be verified and marked.
		❖ the wiring of the CT'S secondary box and the fuses provided shall be checked for continuity.
6.11 6.12		❖ the compression joint type terminal connector suitable for ACSR conductor for CT and PT be provided. Bolted type are also welcome.
		❖ THERE SHOULD BE A PROPER LAYOUT DRAWING AND ROUTE FOR ALL TYPES OF CABLES, TO BE LAID IN THE PLANT. THE POWER CABLES, BOTH HT AND LT SHOULD BE LAID IN SEPARATE ALLIGNMENT, MAY BE SIDE BY SIDE, WITH A GAP AND SEPARATION WITH SAND/BRICKS.
		❖ SCADA CABLES TOO SHOULD BE LAID ALONG WITH THE TOTAL PLANT CABLING OF COURSE IN SEPARATE DUCTS WITH PROPER GAP.
		❖ Route marking of cable routes should be done physically on the entire plant, simultaneously with the cable laying
		❖ the cables are laid as per the approved drawing/specs/cable schedule
	Power and Control Cable Laying	❖ the routing of the cables as per Drawing layout
		❖ that there is no visible physical damage of the cable
6.13		❖ the insulation resistance of the cables before laying
		❖ Cable tray shall not have any sharp edges & if found any it should be grinded
		❖ Appropriate extra length for looping at the two ends shall be allowed
		❖ Cable drum shall be shifted to work place with the help of

		hydraulic cranes
6.14	**Plant Earthing**	❖ **THERE WILL BE TWO EARTHING SYSTEMS-ONE FOR THE DC SYSTEM INCLUDING MODULE AREA AND THE OTHER WILL BE FOR AC SYSTEM INCLUDING INVERTERS AND TRANSFORMERS IN THE SMALL SWITCHYARD.** **It will be better if the two systems are kept separate so that AC Grid earth faults do not back charge the earthing system of DC, raising its voltage, leading to damages to the electronic cards of the DC equipments and SCADA.**
		❖ Main earth mesh will be of 50x6 mm GI strip or 40 mm MS Rod, and interconnections will be with 25X3 mm GI strip.
		❖ if there is any physical damage to the earthing pipes, plates before erecting
		❖ GI flats and rods should be straightened
		❖ Before excavation a total survey to be taken for identifying any already laid line passing through it & the surrounding should be free from obstacles
		❖ The length of pipe electrode shall not be less than 3 Meters. In case hard rock is there then the electrode may be buried inclined to the vertical & the inclinations not more than 30 degree from the vertical
		❖ Earth leads to the electrode shall be laid in a heavy duty GI pipe & connected to the pipe electrode with brass bolts, nuts and washers to be protected against mechanical damage & corrosion by applying bituminous compound at joints
		❖ The soil around the earthing electrode shall be treated to reduce the resistively of the soil by filling the complete depth of electrode with alternative layers of charcoal & salt.
		❖ Earthing conductor around the building shall be buried in earth at a min. distance of 1500mm from outer boundary of the building
		❖ connection between equipment earthing lead & main earthing conductor shall be welded/brazed type
		❖ The earthing of cable tray should be done at a distance not exceeding 10mtrs length by means of minimum size 25x3 MS flat
		❖ Earthing conductor joints shall be double riveted, welded & bolted and 50 mm overlapping is to be maintained at the joints
		❖ Metallic sheath screens & armour of multi core cables shall be earthed at both equipment & switchgear ends.
		❖ **Sheath & armor of single core power cables shall be earthed at switch gear end only**
		❖ All welded joints are painted with black bitumin paint. Check the earth connectivity for each earthing system, & maintain the record.
		❖ The value of the earth resistance at each pit shall be below 1-ohm and for the combined earth mat resistance should not be more than 1- ohms
		❖ the location of earth pits should be as per earthing layout design.
		❖ adequate holes are provided in GI pipes for watering the earth pit in dry months..

		❖ for the earthing conductor buried inside the ground the joints will be welded/riveted & painted with black bitumen paints, thickly coated with bitumen compound to prevent corrosion.
		❖ the layers of charcoal/ salt used for filling as per spec / drawing.
		❖ the brick chamber construction as per the drawing & specifications be done so that watering is done as and when required.
		❖ the continuity of earthing throughout the cable tray
		❖ for earth pit identification, numbering is necessary for periodic checking of earthing resistance as per approved earthing layout.

Civil Works

1. Availability of QA& QC manpower and laboratory to be checked before starting of work and then each month.
2. Sampling for testing of building materials, concrete mix design etc as specified the approved drawings from employer
3. Submission of schedule of tests to be done monthly / quarterly.
4. Stacking and storage of construction materials and components at site
5. All bought out items to be procured from the approved vendor and on approval of Quality plans by employer as per inspection Category
6. Submission of list of Bought out items and their vendors for each of the bought out item identified for approval within the period agreed.

EXCAVATION AND FILLING IN FOUNDATION WORKS

1. Random inspection of nature, type of soil/rock before excavations.
2. Initial ground level to be measured before start of excavation.
3. Final shape and Dimensions of excavations shall be measured for the site under scope
4. Final excavation levels to be measured completely for the complete site under scope.
5. Side slope of final excavation will be checked randomly.
6. Standard proctor Test to be conducted for optimum moisture content and max dry density before fill.
7. In case the earth is borrowed then moisture content test of fill shall be inspected as per approved standards.

Degree Of Compaction Of Fill Backfill

1. Dry density by core cutter method for foundation fill/ backfill for one out of 10 foundations for each compacted layer as per standards.
2. Dry density in place by sand displacement method for area filling, one per 100 sqm for each compacted layer.
3. Density index for area filling and foundation fill as per approved standards

MATERIALS

Cement

- Each consignment of cement shall be duly correlated with manufacturers TC. In case the cement is supplied by the contractor one sample from each lot shall be tested for setting time and compressive strength. Acceptance norms shall be as per relevant IS.

Coarse Aggregate/ Fine aggregates- will be checked for
1. Moisture Content 2. Specific Gravity 3. Water Absorption 4. Sieve Analysis 5. Flakeness Index 6. Elongation index 7. Deleterious materials (coal & lignite, clay lumps, material finer than 75 micron sieve, sofi fragment, shale 8. Crushing value abrasion value and impact value as per approved standards by employer.
Water- Test for quality as per approved procedures.
Concrete
1. Test for Crushing strength (works Tests cubes) 2. Workability - slump test 3. Admixtures of approved brand will be used
Concrete conveying, placing & compaction
1. Mixing of concrete shall be done in a approved mixer such as to produce a homogenous mix and it shall be calibrated suitable agreed period. 2. Arrangement for transportation & placement of concrete before clearance. 3. Handling and transportation of concrete as per construction methodology 4. Placement of concrete as per construction methodology 5. Compacting as per IS standard. 6. Curing with period as per IS standard
Test of Settled RCC structures
1. Dimensional check on finished structures & dimensional tolerances 2. Water Tightness Test of liquid retaining structure / tanks
➢ **Reinforcement steel**To be checked for physical and chemical and structural properties as per IS codes for the batch of supply.
Placement of Reinforced steel
1. Bar bending schedule with necessary lap, Spacers & Chairs 2. Bending of bars, cutting tolerance 3. Acceptance - Cover, spacing of bars, spacers and chairs after the reinforcement cage is put inside the framework
Staging and Forms
1. Visual checks of materials and accessories as per standards before start of work. 2. Visual checks for soundness of staging, shuttering and scaffolding including application of mould oil / release agent before start of work. 3. Acceptance of formwork before start of concreting each time.
Grouting
1. Each batch of material to be checked for source from EIC approved list and as per approved specifications. 2. Physical inspection prior to start of work for type of mix - fluid mix, plastic mix, stiff mix etc. 3. Random physical inspection for mixing, placement, application and grout pressure 4. Compressive strength to be checked as per approved procedure as per specification, 5. Acceptance of the grouts by physical inspection of each grout.

BRICK MASONARY

- **Test on Bricks for -**
  Dimensions, shape, compressive strength, water absorption, warpage, efflorescence as per IS standards.
- **Masonry construction-**
  Physical test for workmanship, verticality and alignment for 100% technical specification,

Plastering Workmanship

1. Curing to be checked 100% to be as per approved schedule.
2. Thickness and finishing of plaster, grooves etc by visual inspection and measurements
3. Trueness of plastering system as per approved specifications and standards

Water Proofing

1. Methodology to be as per approved technical specifications
2. Polyurethane based coating, polyester scrim cloth, extruded HD dimpled polyurethane with materials from EIC approved source and as per tech specification.

Water supply fittings and fixtures

1. Material- GI/ MS pipes and fittings
2. Disinfection- Before use
3. Hydraulic test- Before use / Leakage
4. Water Storage Tanks- As per approved drawing and of the declared capacity required for cleaning the modules at the place specified in final approved layout of the plant.

Earthing Mat (Grounding System)

- **Material:**

1. Earthing mat- Testing of each lot with laying as per approved drawings and calculations with materials from approved sources.
2. Weld sizes & length- Inspected visually and with specifications as per approved drawings.
3. DP Test- conducted for 10% of the lot
4. Earth test- As per approved procedure for the approved design of earth mat

STRUCTURAL STEEL MATERIAL (For Site Fabrication)

Pre-welding requirements

1. Welding procedure specification as per the employer approved procedures.
2. Welder to be having suitable qualification as agreed between employer and bidder.
3. Welding consumables as per approved specs and makes.

FIT-UP

1. Marking and Cutting as per measured details of the site to comply with the laid down approved specification.
2. Match markings for trial assembled components. For each fit-up.
3. Weld fit-up- Edge preparation/ Gap alignment with stiffeners as required

Welding requirements

1. Sequence of welding to be inspected for as per standards
2. Removal/ grinding of temporal attachments to be inspected to be for as per standards.
3. Completeness after welding-Dimensions/ distortion to be physically inspected.
4. Completeness of welding (each butt &fillet weld) as per standards.

Fillet Welds

i. size and visual examination
ii. Dye Penetration Test

Butt Welds

i. Visual examination
ii. DPT- 100% Physical inspection on all butt welds after back gauging on root run. All to be back gauged before DTP.
iii. Radiography Test as per standards

Pre-assembly and erection checks

1. Punch Erection marks and match marks on members
2. Pre-assembly as per match mark.
3. Camber, sweep and total length after trial assembly of structure
4. Control assembly check at shop
5. Alignment, slopes, level, tolerances of erected member
6. Tightening of bolts/ Torque including foundation bolts with lock nuts
7. Completion of all erection fillet & butt welds. Completion of primer & intermediate coat of paint.
8. Acceptance of erected structure.

ROAD WORKS

Construction of Sub-Grade and earthen/hard shoulders-

1. Standard Procter test for One in every 2000 cum for each type and source of fill materials
2. Moisture content of fill before compaction for one in every 2000 cum for each type and source of fill materials.
3. Dry density test by core cutter method
4. Check for line, grade and cross section for one in every 500 Sq. m. area.

Water Bound Macadam (Non-Bituminous) for base course and sub-base course

1. Aggregate impact value by the named instrument with test of one in every 1000 cum of test aggregate and as per approved technical specification
2. Grading for one in 1000 cum of aggregate with specification as per standards and approved procedures.
3. Flakiness index and elongation index with test on one per 1000 cum of aggregate
4. Atterberg Limits of binding material with test on one per 1000 cum of aggregate
5. Atterberg Limits of portion of aggregate with test on one per 1000 cum of aggregate

Bituminous Macadam for base and binder course

1. Quality of binder as per IS standards
2. Aggregate impact Value / Los angles with test on one per source and as per IS standards.
3. Flakiness index and elongation index as per IS standards and test on one per 100 cum of aggregate.
4. Stripping value of aggregate (immersion tray test) s per standards.
5. Water sensitivity of mix.
6. Grading of aggregates
7. Water absorption of aggregate
8. Soundness (Magnesium and Sodium Sulphate)
9. Percentage of fractured faces
10. Binder content and aggregate grading

11. Control of Temperature of binder and aggregate for mixing and of the mix at the time of laying and rolling
12. Rate of spread of mixed materials
13. Density of compacted Layer

Bituminous Surfacing - Openradedoremixcaroet and Seal coat

1. Quality of binder
2. Aggregate impact Value / Los angles abrasion value
3. Flakiness index and elongation index of aggregates
4. Stripping value of aggregate (immersion tray test)
5. Water absorption test
6. Water sensitivity of mix
7. Grading of aggregates
8. Soundness (Magnesium and Sodium sulphate)
9. Polished stone value
10. Temperature of binder at application
11. Binder content

RCC Pavements

- Concrete - Material, Mix design, Production, Transportation, Placement, Compaction, Curing, Test on green concrete, Test on hardened concrete etc.

Alignment. Level. Surface regularity and rectification

1. Horizontal alignment, Surface levels and Surface regularity
2. Rectification

APPROVALS

PRE-COMMISSIONING CHECKS

COMMISSIONING

&

FINAL CHARGING

Of

MW CAPACITY PV SOLAR POWER PROJECT

Commissioning Engineers must know that-;

"The electricity is not visible"

"It does not give a second chance to the operator"

1.0 Plant Commissioning

The site engineers who are associated with the construction and charging of any power project should take following precautions while working on the equipments-
They should have line tester, multi-meter, Insulating shoes and other safety equipments handy with them as Electricity is neither visible nor does it give you a second chance.
The Solar cell-the moment it is exposed to Sun- starts building voltage and the string will have a good level of it.
This document has three sections-
Section - 1 Registration and Approvals
Section - 2 Technical
Section - 3 Format for Technical parameters

Section 1

1.1 PRIOR TO COMMISSIONIG THE PLANT, THE RELATED DOCUMENTS, SHOULD BE CHECKED TO CONFIRM THAT PLANT FULFILLS THE REQUIREMENTS AS ENVISAGED BY ALL AUTHORITIES, SUCH AS-:

1. RREC- Plant is registered with RREC and has relevant approval of the project.
2. DISCOMS- PPA is signed with the Discoms for purchase of power generated.
3. MNRE guide lines, are strictly followed.
4. Metering as per norms is provided.

Commissioning of a solar photovoltaic power plant may be defined as:-A solar PV project will be considered as commissioned if energy has flown into the grid at rated capacity of the plant, taking into account the availability of solar radiation on that day.
For above, the following documentation are required:

1.2. A. Installation report: This will include the following information:

1	Capacity of the plant (MW)	
2	Technology used (Mono/Multi Crystalline/Thin Film/Others; please specify Along with capacities of each type)	
3	Rating of the each module (Wp)	
4	Angle from horizontal at which array is installed	
5	Number of modules installed of each type	
6	Source(s) of the modules(Name of supplier with address)	
7	Number of PCUs installed	
8	Source of the PCUs (Name of Supplier with address)	

9	Date of installation of Full capacity of PV array PCUs Transformer	
10	Date of inter-connection with the grid	
Please attach the photographs taken from various angles covering installation of important components of the solar power plant		

B. Commissioning Report:

This will include the following information:

i) A certificate of the Discom/STU stating that the plant has been synchronized with the grid

ii) Data on generation of the plant on the day of commissioning along with available solar radiation on the plane of the PV array

iii) Data on peak generation reached during the day of commissioning

iv) Two different snap shots of the meter reading before and after commissioning during the day

v) A justification for the lower peak generation than the rated capacity (if it is so) supported by the measured data of solar radiation during the corresponding time period

vi) A certificate issued by s Committee, authorized by NVVN, stating that the date of commissioning

vii) Photographs of the project from different angles.

C. CEIG- SAFETY GUIDELINES-

- The plant should have CEIG approval to energize the plant as a whole and the associated transmission line.

D. REVENUE DEPTT.

- Plant is installed on the land which is in the name of the plant owner and it is approachable by road.
- Application for CLU has been submitted to the collector Jodhpur, with deposition of fee.

E. ENERGY EXPORT METERING

- The Metering equipment at the Delivery Point is in accordance with relevant provisions of Metering Code as applicable for generating stations/CPP and is provided by the Power Producer himself at his cost. RVPN/ concerned Discom will seal the meters and metering boxes.
- The injection and metering arrangement is finalized by RVPN / Discom in consultation with RREC on the basis of details furnished by Power Producer. The Power Producer or RVPN/Discom shall be responsible for security & protection of metering arrangement based on the location of metering arrangement as stipulated.

PCB clearance

- Clearance to establish the project has been obtained.
- Registration with labor department is done.

APART FROM ABOVE, THE FOLLOWING ASPECTS IN GENERAL BE CHECKED AND RECORDED

- Please take a round of the plant and ensure that the plant boundary wall is complete in all respect and there are no chances of any accidental entry of villagers and animals in the plant area.
- Please ensure that the plant area is safe/leveled/and has pathways for movement of employees and the vehicles.
- First aid facility is available in the control room.
- Fire fighting equipments are available in the plant.
- Toilet facility and drinking water facility is available.

- The route of cable is marked by metallic indicators.
- Safety equipments are available at site

Section - 2

2.1 PRE-COMMISSIONING CHECKS

- Starting from solar cells and going to power evacuation DP, you should collect all the test results, carried out by the contractors and supplier of each of the equipment, whether in the manufacturing units or at site. Read them carefully and note the-
- Design parameters
- Protection settings
- Installation guidelines
- Precautions during handling
- Safety measures
- Analyze and follow them
- Check availability of 110 or 24 volt DC supply in HT Panels and any other system, wherever required

Solar Modules –

- Check that polarity of the cells, forming strings, is correctly connected.
- The positive cable is marked with red tape and negative with green tape at the end.
- DC SYSTEM is Floating or unearthed for maximum reliability and safety
- Since the sun is available, safely measure the output voltage of each string with multi-meter at SCB terminals- it will be in the range of 600 to 700 Volts.
- The module mounting structure is properly earthed as per the design.

2.2 String Combiner Box –

- Check HRC fuse rating it should be equal to the DC input current + 2 amps.
- Check settings of over current, earth fault, over voltage, under voltage and short circuit protections of the MCB and record them.
- Surge protection device should be properly earthed at one end.
- Check SCADA outputs are available along with supply.

2.3 Cables – (Common for entire plant cables)

- Check that cables are properly laid and dressing done by laces.
- Ferruling and tag marking to identify each cable is provided.
- The cable lugs are properly crimped and the glands are in place.
- Each cable should have been tested for IR value by meggar before connecting.
- Armoring is earthed wherever required, and taped, if earthing is not required.

2.4 Inverter –

- Check polarity of inverter for DC and confirm that +ve of 120/150 Sq.mm cable from SCB is connected to +ve port and –ve to –ve port.
- Check the settings of over voltage, over current, earth fault, reverse power, short circuit and any other protection both on DC & AC side of the inverter
- Check- HRC fuses ratings on both +ve and –ve side.
- The manufacturer's supervisory engineer's report copy be taken and studied.

- Each inverter is earthed at two points.
- Cooling fan units are properly mounted and ready to run.

2.5 Power Transformers –

- Check break down voltage (BDV) of the oil, it should be above 70KV at 4 mm gap.
- The moisture content in the oil should be less than 20 PPM.
- Check IR values (with meggar) of HV and LV windings, it should be near infinity.
- Oil level in the transformer should be above the mark in MOG.
- Each transformer is earthed at two points.
- HV and LV cables are properly tightened to the respective bushings.
- Transformer protections - Bucholtz, OTI, WTI, PRV and low oil level (Alarm & Trip) are properly connected.
- The OCTC indicator should be at normal tap position with locking arrangement.
- The vector grouping test done by suppliers engineer should be checked to confirm that the transformer is YnD11.

2.6 33 KV VCB-

- See that all meters – Ammeter, voltmeter, Power factor Meter, MFM and Indication lamps- are provided on the front of the panel and are clearly visible to the operator
- All the relays are in position with the relay settings indicated on each relay.
- Check the availability of protection supply to each panel.
- Check the availability of 220V AC auxiliary supply to each panel.
- Each panel is properly earthed.
- Check the operation of spring charging mechanism both electrically and manually.
- Select the breaker position on test and operate- on and off -the breaker through push button and also by each of the protective relays.
- Take the IR value of all the three phases with cable connection and without cable connection.
- Carry out the testing of the relays; arrange the relay settings of all the relays on each of the HT panel.
- Check the protection philosophy by opening of the individual breaker by simulating the relays- over current, over voltage, under voltage, earth fault and short circuit etc.

2.7 HT Cables –

- Check the IR value of every section of the cable and record.
- See that the underground cable route is marked.
- There should be identification nos. on each laid cable

2.8 Four/Double Pole structure–

- Physically inspect the DP/Four Pole and the area and see that jumpers from 33 kv line to LA to Isolator to Drop-Out is done properly.
- Open and close the isolator on off load.
- Check the DO fuse rating.

- Check the CT-PT connections and ensure that the primary of the PT is solidly earthed and no secondary of the CT is open circuit.
- MFM and TVM is connected to both CT and PT to give the readings of –power (KW), voltage (volts), current (Amp.), Power factor, KVARH and KWH etc.

2.9 Transmission Line –

- Inspect the132 or 66 or 33 KV line and see that conductor spacing is visually ok on the entire route. Visibly observe that the clearances are maintained as per norms
- The stays are properly tightened with insulator in between.

2.10 Plant Earthing and Lightening Protections-

- Measure total plant earth resistance, it should be less than 1 ohm.
- Check healthiness of the system by measuring earthing resistance at all corners and near all buildings. Earth pits are covered.

2.11 ACDB

- Check availability of ACDB with proper switches and fuses both at input and output.
- The Aux. transformer should be ready to take 33 kv supply and feed the aux. power supply system at 415 volt
- Once an overall checking of the entire system, as detailed above is, completed and recorded in the check register, the next step is commissioning.

At this point, every dis-connecter, as identified below, will remain in OPEN position. These are to be PUT-ON step by step, to energize the concerning loop, after taking into considerations the steps as explained here below

- 33 kv line isolator
- 33 kv breakers of all HT panels.
- ACDB outgoing switches
- ACDB I/C and O/G switches
- SCB I/C and O/G switches
- Inverter I/C and O/G switches

3. COMMISSIONG AND FINAL CHARGING STEPS

The commissioning activity is divided into the following groups

3.1 Physically inspect the total plant and ensure that the complete area and circuitry is in its best condition and all men and materials have been removed from the plant equipment area.

3.2 Charging the 33kv transmission line from GSS to plant DP.

- Meggar the transmission line with a 5kv meggar, the IR value should be infinity or more than 500 k-ohms.
- Keep the DP line isolator open and give clearance for line charging from GSS end.
- Once the line is charged from GSS end, close the isolator so that the 33kv supply reaches up to the VCB.

3.3 Charging the 33 kv HT Panels

- Once the 33kv line is charged, it is ready to back charge the transformers.

3.4 Charging the 33/0.415kv auxiliary transformer and the ACDB-

- Charge the 33kv/415 volt auxiliary transformer by closing the 33kv side breaker.

- Check, by multi-meter, that 415V LT 3phase supply is available at ACDB inlet by checking/ measuring the voltages in RY-YB-BR-RN-YN-BN. Value should be 415V phase to phase and 220V phase to neutral.
- Close the ACDB breaker to charge the 415 LT bus with all out going switches in open condition. These switches can be closed as and when the LT supply is required in different areas as per the relevant drawing.

3.5 Charging 33/.300 kv or any capacity power transformers-

- Take clearance of the equipment and inverter supplier for charging the power transformers from 33kv side.
- Close the 33kv breaker of the respective HT panel to charge the transformer.
- Check the relay settings of the
- The 300V supply is now available up-to the the inverter
- Inform the inverter engineer that this 300V LT supply can be given for synchronization of the two supplies -the solar supply and the grid supply.

Now Grid AC supply stands extended up-to output MCB of the inverter, ready to synchronize with solar power coming from other end, of course, inverted thro inverter. After reaching up-to this step, we have to start charging the solar end supply in steps as detailed below.

3.6 Charging the Solar Cells-

- Actually the solar cells are automatically charged as soon as they face the sun.
- String formation-22 cells per string-should be checked to confirm the Voc-it should be as per our designed value.
- Check that each string is correctly made and the polarity is also correct.
- Insert 15 amp fuses in both +ve and –ve circuits in the SCB. The solar supply will reach up-to the output of SCB
- Take clearance from inverter supplier for closing the output MCB of the SCB.
- After clearance from inverter, this MCB can be closed, which will extend the solar supply up to DC Box of inverter.
- Check that the inverter engineer has tested and set the values of the protection relay settings.

3.7 Charging the inverter-

- Check, by simulation, that the protection relays are actually tripping the system on occurrence of faults in the loop.
- Close the DC Box output MCB to charge the inverter.
- Check that the output from the inverter at output MCB of inverter is 300 volt, 3-phase, 50 c/s A.C. supply.

At this point the DC Power-inverted to AC Power- has extended up-to the other end of the Inverter output MCB. After both the supplies-AC and Inverted DC- are available across the MCB, the synchronization of the two supplies can be resorted to on auto.

3.8 FORMAT FOR COMMISSIONING CHECKS

S.No	Name of Equipment	Parameter	Unit	Design/ rescribe Value	Actual value as measured	Setting/Measring Agency	Measuring Equipments
1,	Solar Module-Random Sample	Voc	Volts				Multimeter
		Vmpp	Volts				Multimeter
		Isc	Amps.				Onload by Tongue Tester
		Impp	Amps.				Onload by Tongue Teste
2	String Combiner Bo Random strings	Input String Voltage	Volts				
		Input String Current	Amps.				
		SCB output Voltage	Volts				
		SCB Output Current	Amps.				
		Fuse Rating +Ve& -Ve	Amps.				
		O/C Setting	Amps.				
		E/F Setting	Amps.				
		O/V Setting	Volts				
3	DC Combiner Box	Inputs	Nos				
		Fuse Rating	Amps.				
		O/C Setting	Amps.				
		E/F Setting	Amps.				
		O/V Setting	Volts				
4	Inverter	DCinput Volt.	Volts				
		DCinput Current	Amps.				
		ACOutput Vol	Volts				
		AC Output Current	Amps.				
		O/C Setting	Amps.				
		O/V Setting	Volts				
		U/V Setting	Volts				
		Temperature	Deg.Cent				

		E/F Setting	Amps.				
5	L.T ACB	Manual Operation Off Loa Condition	Locked After Closing/ Opening				
6	Transformer	IR Value LT	Mega Ohm				Meggar
		IR Value HT	Mega Ohm				Meggar
		BDV Oil	Kv				Oil testing set
		Oil Level in Conservator	OK				Visual
		Bucholtz Alarm	Available				Visual
		Bucholtz Trip	Available				Visual
		OTI Alarm	Degree Celsius				Gauge
		OTI trip	Degree Celsius				Gauge
		WTI Alarm	Degree Celsius				Gauge
		WTI Trip	Degree Celsius				Gauge
7	3 Kv I/C HT Panels- 1,2,3,4,5	O/C Setting	Amps.				
		O/V Setting	Volts				
		E/F Setting	Amps.				
		U/V Setting	Volts				
8	3 Kv P/E HT Panel Aux. HT Panel	O/C Setting	Amps.				
		O/V Setting	Volts				
		E/F Setting	Amps.				
		U/V Setting	Volts				
		O/C Setting	Amps.				
		O/V Setting	Volts				
		E/F Setting	Amps.				
		U/V Setting	Volts				
9	Power Evacuation DP Instant Readings	Fuse Rating	Amps.				
		Voltage	KV				
		Kw	KW				
		Kwh	Kwh				
		Kvarh	Kvarh				

		Power Factor					
		PT Voltage					
		CT Current					
10	33 kv line	IR value	M-Ohm				
11	TVM	All readings	KW				
			Kwh				
			Kvarh				
		Phase Rotation	Clockwise /Anticlockwise				
12	Aux. Transformer	IR Value LT	Mega Ohm				
		IR Value HT	Mega Ohm				
		BDV Oil	Kv				
		Oil Level in Conservator	OK				

THE END.....

OPERATION and MAINTAINANCE
OF
MW CAPACITY GRID TIED SOLAR PV POWER PROJECTS

[This section defines the operation and maintenance requirements for all solar power plants installed in any area. Even though Solar PV power generating plants require low maintenance and servicing, as compared to most of other different power generating technologies, however, suitable maintenance of a PV plant is essential to optimize energy yield and maximize the life of the system.]

The PV solar power plant is a static generator, which does not have any movable part in its entire system. Also the power generation automatically starts when the sun rays fall on the solar modules. The inverter output power gets automatically synchronized with the grid power and power starts flowing to the grid. This process automatically stops as and when the sun stops shining on the modules. Therefore, there is practically no need for any operation engineer for starting/regulating/stopping the power generation process. Only one engineer can walk around the plant and see that above process has started, he can take readings from the provided meters.

If SCADA is available, then this one engineer can see the complete plant generation process from the control room itself.

For maintenance also, not much of staff is required, provided the plant erection work is done in the best of condition. The cabling and cable terminations of the strings and DC and AC power circuits are one field which needs maximum attention. The module mounting structure integrity is the second major aspect especially in the rainy and stormy season. The third aspect is module cleaning frequency and the system/method adopted for this. It has to be done once in a month. Therefore here again not much of staff is required for maintenance of solar power plants.

But each and every equipment of the plant, such as Solar Modules, String Combiner box, Inverters, Transformers, HT and LT Switchgear and Cables have a periodic check technical procedure to be followed on Daily/Weekly/Monthly/Quarterly/Six Monthly/Yearly basis.

Finally the plant owners also require a daily data feedback about the generation level; damages of the plant due to various reasons, grid outage time monitoring and equipment's health status. Hence proper maintenance engineers are required for this part of the work.

Therefore a small team of "The Operation and Maintenance" engineers and technicians is required to look after the commissioned plants, and see that these plants are giving the maximum output. This team should devote itself to increase the generation only and desist from finding faults in the design.

"ONLY AUTHORISED PERSONS SHOULD BE ALLOWED TO WORK ON THE LIVE EQUIPMENTS"

1. Engineers and staff required for the works of O&M

To carry out this work, a team of the following engineers and technicians is required for a 5-mw plant

1. Engineer in-charge O&M---------1
2. Assistant O&M engineer------1
3. Dip. Engineer electrical--------2
4. ITI electrician---------------------2
5. Water washing contractor and manual labor on day to day need basis.
6. The number of persons required depends upon the size of the plant-bigger the plant capacity more will be the number.

A DAILY MONITORING CHART OF GENERATION IS TO BE MAINTAINED FOR ALL PLANTS AND DAY TO DAY GENERATION NEEDS TO BE RECORDED AND COMPARED FOR ANALYSIS BY HEAD OFFICE AND TECHNICAL WING.

2. SCOPE OF WORK-

- The aim is to operate and maintain the plant in a healthy condition, there-by ensuring highest levels of efficiency, maximum power generation and also to prevent possible malfunctions in the plant.
- To arrange guide lines and recommendations ofequipment suppliers about -method, frequency, safety of all equipments and operation and maintenance procedures.
- The guidelines should be readily available with the O&M engineers in both soft and hard copy.
- Deployment of engineering personnel, technicians and support staff for operation, maintenance and attending to system faults.
- Arrange day-to-day operation of the plant to achieve maximum power generation.
- Maintain daily records of all parameters and happenings at the plant.
- Arrange preventive maintenance of the plant.
- Arrange major maintenance in consultation with the owner and the manufactures of the equipment.
- To prepare a list of spares to be arranged by power plant owner. Maintain a stock of necessary spares based on recommendation of equipment supplier and our own past experience. Keep special tools for operation and maintenance.
- Visual inspection of the general state of the plant and solution of any problems found or, if not possible immediately, planning of attending to problem/minor issue found, in order to solve them in next visit.
- Check warnings and take the necessary corrective and/or preventive measures, that will reduce the chances of any electricity relate default by strictly following the safety measure so that no accident takes place at the plant.
- Analyze solar irradiance data and other weather parameters and correlate them with the quantum of energy generated daily.
- Use safety tools and keep fire incidences to level zero.
- Location of the project should be known to the O&M engineers for obvious reasons-State-District-Tehsil-Village-Latitude-Longitude-GHI-Temperatures-Humidity levels-Rainfall data-Wind Velocity etc.

3.0 OPERATION-PERFORMANCE-MONITORING-EVALUATION

The operation of the plant starts one hour before the sunrise and continues till one hour after the sunset. To optimize system performance, there is a need to ensure that the plant components function efficiently throughout the lifetime of the plant. Continuous monitoring of PV systems is essential to maximize the availability and yield of the system.

The work involves the following steps

- Regular reading of units generated should be recorded everyday and this should be intimated to all concerned. A daily analysis is necessary to compare the generation with Sun and Grid availability and reasons for low generation intraday must be identified.

- The solar radiation and wind speed needs to be monitored daily.
- A SCADA system is able to monitor the real-time efficiency, indicating each and every parameter such as DC power input-AC power output-Voltages-Currents-Weather reports etc. The O&M engineers must continuously compare it with the theoretical efficiency of the plant, to assess if the system is operating optimally. This information can be used by the O&M team to establish the general condition of the system and schedule urgent repair or maintenance activities such as cleaning, inverter room temperatures and air flows.

4. CHECK LIST FOR DAILY/ROUTINE OPERATION

This is a job of 6.00AM to 7.00PM, i.e. when the sun is available. For explaining O&M procedure, we start from the modules and go to the power evacuation feeder.-;

- At 6.00 AM, the team has to ensure that each inverter is getting the grid supply and is ready to synchronize with the grid power, as and when the sun rises and modules start generation. Once the generation starts, the team should personally check that, each equipment of the plant is working properly.
- Once the generation starts, it is the duty of this team to record the values of-Voltage, Current, Power, Power factor, Temperatures, Energy output and any other parameter as indicated in the C&R panels and the panels of all equipments. This should be recorded in the daily log register.
- Any abnormality noticed, should first be recorded in the O&M register, checked properly and discussed with the team members, and then action be taken to rectify the fault. Special watch for temperature rise of modules, cables, inverters and transformers is an important aspect as the temperature is directly related to the efficiency of the concerned equipment. Note should be made about the necessity of water washing of the modules and planning be done to do it as per the plan.
- All safety tools should be available with the team members and these should be correctly used to work on the live parts.

5.0 Plant Maintenance

Maintenance can be divided into following groups-

(A) Scheduled or Preventive or Planned Maintenance

(B) Unscheduled or Fault Maintenance

A. Scheduled or preventive or planned maintenance

The maintenance team should make plan of the maintenance activity well in advance. This plan has to be made after checking the condition of the complete system and all components of the plant. This plan should also be based on the maintenance guideline as given by the equipment suppliers and also the past experience of the maintenance engineers. Once this plan is made, the work should be carried out during no-sun hours or during planned shut downs. It will reduce the number of faults from occurring, as well as to keep the plant operating at its optimum level. It includes the following works-:

i) Module cleaning.

ii) Checking module connection integrity.

iii) Checking string combiner boxes.

iv) Inverter servicing.

v) Inspecting mechanical integrity of mounting structures.

vi) Vegetation control.

viii) Routine balance of plant servicing/inspection of site substation and four pole structure.

ix) Site fencing and security system.

x) Site access routes and site roads.

xi). Site buildings and containers.

xii) Maintenance of fire-fighting equipment.

- Although scheduled maintenance will both maximize production and prolong the life of the plant, it does represent a cost to the project. Over doing the maintenance in this type of static generator is a waste of time, money and resources. **Therefore, the aim should be to seek the optimum balance between cost of scheduled maintenance and increased generation through out the life of the system.**

Specific scheduled maintenance tasks, equipment wise, are covered in the following sections.

a. Module Cleaning

- If the cost to clean the modules is less than the increased revenue then it is beneficial to clean the modules. Module cleaning is simple but important task. It can produce significant and immediate benefits in terms of energy yield. The optimum frequency of module cleaning can be determined by assessing the costs and benefits of conducting the procedure.
- The benefit of cleaning should be seen in an improved performance ratio due to the lower soiling loss—and result an increase in revenue. A cost estimate to clean all the modules at the PV plant should be obtained locally.
- Daily, visually check that the modules are clean-if not; wash them with water in the evening. The frequency of module cleaning will depend on local site conditions, like prevalence of dust or rain and the time of year. The duration between clean-ups is likely to vary with weather patterns i.e., cleaning during rainy periods is less likely to be required. There should be a pre-planned module cleaning cycle, which should be followed and updated according to the dirt settlement on the modules.
- Hot modules should never be water washed. Thin dust does not affect the generation. However, it is generally recommended to clean the modules at least once monthly.
- Uncalled for and more water washing is to be strictly avoided-as it damages the modules and water is wasted. If the system efficiency is found to be below the expected efficiency, then module cleaning should be scheduled as necessary.

b. Module Connection Integrity/String Monitoring

- Checking module connection integrity is important. This should be done on daily basis. The operation engineer should make plan to monitor a minimum number of strings daily. He can monitor all the strings of two SCBs daily and should continue the cycle till all strings are checked. Then repeat the cycle so that total plant is visited twice in a month.
- For carrying out the string monitoring, you need a hand held DC meter which gives the current flowing in the string. Currents in the strings should be equal for similar strings during peak load hour-i.e.12 noon to 2PM. Therefore, the +ve and–ve connections between modules, within each string should be checked periodically, at least once in a month.
- During string monitoring, please check that, every terminal connector is properly connected.
- Check that there is no heating and sparking in the MC-4 connecters and the terminations. In such cases, faults within each string of modules may be detected.

c. String Combiner Box

- All string combiner boxes should be checked periodically for water ingress, dirt or dust accumulation and integrity of the connections within the boxes. Loose connections could affect the over all performance of the PV plant. Any accumulation of water, dirt or dust could cause corrosion or short circuit within the junction box.
- Practically these string combiner boxes collect DC voltage and current at one point from all strings. Check that each string is delivering designed voltage and current. This checking is a regular feature and each string should be checked once in 15days. Record the values in a register and pass it on to all concerned.
- If, on comparison, you find the values less in any string, then you must find out the reasons and rectify the fault. Chances are-Either+/-fuses are blown, or the terminal connector is loose, or cable is damaged. Periodic checks on the integrity of the fuses in the string combiner boxes should be conducted.

. Inverter Servicing

- Generally, inverter faults are the most common cause of system downtime in PV power plants. Therefore, the scheduled maintenance of inverters should be treated as a centrally important part of the O&M strategy. Close and periodical monitoring of each and every inverter is necessary.
- Temperature monitoring should be given more attention. The permissible range is maximum- 50 deg.cent. Inside the Inverter panel and40 deg.in the room. In summer, please see that the cooling fans tarts on auto and the thermostat setting is as prescribed-50deg.Cent.
- Check all the parameters that are available on the panel viz. Voltage, Current, Power, Energy etc. The values should be as prescribed.

- Cleanliness and sealing of the inverter room is very important and it should be ensured that there are no chances of entry of mosquitoes and reptiles in the room. All the equipments of the Inverter should be properly tightened
- The daily/weekly/quarterly/six monthly/annual preventative maintenance for an inverter should, as a minimum, include the following: Visual inspections, Cleaning/replacing cooling fan filters, removal of dust from electronic components, tightening of any loose connections and any other additional analysis and diagnostics recommended by the manufacturer.

e. Structural Integrity

- The module mounting assembly, cable conduits and any other structures, built for the PV plant, should be checked periodically for mechanical integrity and signs of corrosion. Check that modules are properly secured and held with tightened nuts and bolts.
- This will include an inspection of support structure foundations for evidence of erosion from water run-off. Special attention should be paid to the structures during rains, storms and high velocity winds. The remaining systems within a PV power plant, including the monitoring and security systems, Transformer, RMU, HT System, auxiliary power supplies, and communication systems should be checked and serviced regularly.

f. Transformer:

- Check that there is no leakage of oil from any joint of the transformer. The oil level in the conservator is as per the marked level. If less, it is to be filled during off load time, after taking shutdown of the circuit .Check the OTI, WTI, PRV and Bucholtz relay and record the temperatures at peak load and at maximum sun time.
- Check IR values of the HV and LV with respect to each other and earth and it should not show a sign of reduction from the last taken values
  - Oil filtration after winter/after rains is required if its BDV goes down below 70 kv for 1-minute.

g. HT Switch gear with Isolators and associated HT and LT cables

- The HT panels/Breakers should be in a very healthy state, because they save the plant from damages during electrical faults. Visually check them for any fault or defect and note down, for attending to fault at the time of shutdown. Check and confirm that all breaker trip protections are in circuit. The VCB normally does not require any daily/weekly/monthly maintenance, but after every tripping on fault a minute visual inspection is required.

h. Earthling System

- The plant earthing network is a silent protector; it should give a resistance value of less than

- 1- ohm. In case the value is going high, the net work needs to be augmented and watering of the earth pits should be done at least once a week. During peak summer the earth pits should be watered once in 15 days.

There should be less than 1-Volt potential between the earthing mesh and the neutral of the 415 volt. grid auxiliary supply. If it is more please improve the earthing by adding more earth electrodes.

Vegetation Control

- Vegetation control and ground keeping are important scheduled tasks for solar PV power plants, since there is a strong likelihood for vegetation (for example, long grass, trees or shrubs) to shade the modules. The ground keeping can also reduce the risk of soiling (from leaves, pollen or dust) on the modules.Dense vegetation becomes a resting place for reptiles-like snakes and lizards.

6.0 **Unscheduled Maintenance or Fault attending, after the fault has occurred and damage has taken place.**

This is carried out in response to failures. Common unscheduled maintenance requirements includes the following

a.) Tightening cable connections that have loosened.

b.) Replacing blown fuses.

c.) Repairing lightning damage.

d.) Repairing equipment damaged by intruders or during module cleaning.

e.) Rectifying supervisory control and data acquisition (SCADA)faults.

f.) Repairing mounting structure faults.

Above are further detailed as below

- Unscheduled maintenance is carried out in response to failures. As such, the key parameter,
- when considering unscheduled maintenance is-Diagnosis / Speed of response / Repair time.

- Although, the shortest possible response is preferable to avoid any effect on power generation and maintain the energy yield. There should be minimum response time and quick actions should be taken against such failures, resulting in repair or replacement of faulty equipment depending on the type of faults.

- The majority of unscheduled maintenance issues are related to the inverters. This can be attributed to their complex internal electronics, which are under constant operation

7.0 Spares Managements-

In order to facilitate a rapid response, a suitably stocked spares inventory is essential. The numbers of spares required will depend on the size of the plant and site-specific parameters. Adequate supplies of the following components should be held:

- Mounting structure pieces.

- Spare parts of String combiner box like Fuses etc.
- DC and AC cabling components.
- Communications equipment.
- Modules
- Spare of inverters

It is important that spares stock levels are maintained. Therefore, when the O&M team uses some spares they should replenish the stocks as soon as possible. This arrangement will reduce the time gap between the identification of the fault and replacement of the non-operational component.

Consultation with manufacturers to detail the spare parts inventory, based upon estimated component life times and failure rates, is recommended.

8.TOOLS REQUIRED FOR OPERATION AND MAINTENANCE

a. Digital Multi Meter Make Fluke 15B -2No 4000 count digital millimeter- Voltage measurement-1000VAC/DC, Current measurement-10 AAC/DC CAT-II

b.) WACO0-5kVInsulationMeggar- Zero to Infinity M-Ohm

c.) Earth Meggar- 3/4 Pin Method

c.) Digital Clamp meter- For both AC &DC Monitoring

d.) Electrical Tool Kit with selected tools

9.0 Daily operation and maintenance log sheets-; The O&M site engineers will have to fill the following log sheets daily. They must prepare the template in the computer itself so that the daily observations are sent to the company heads and Directors. The format of the log sheets are as below-;

SHEET-1

Daily Generation Summary sheet no 1. NAME OF THE PLANT....................
MW/KW capacity of the plant.....................Month...

Date	Units in Inverter meter	Units in ABT meter	Inveter to ABT meter Loss (%)	*Net Operation Hrs. of Plant*	*Total grid fail Hrs. (hh:mm:ss)*	*Avg. Radiation (KWh/ m²)*	*Avg. Ambient temp.* (°C)	*Weather Information*	*Rain duration (hh:mm:ss)*	*Remarks*
									_	If number of inverters in one plant are more than one, then the inverter energy units

										column can be added there itself.

SHEET NO-2

Date	**Daily total plant generation sheet no.2 Name of Plant…………………..Plant Capacity…………..Month……………..**							
	Inv-1			*Inv-2*			*Total plant daily generation Total units of the day*	*Add similar table for other inverters*
	Opening Reading	Closing Reading	Total	Opening Reading	Closing Reading	Total		

Total								

Module cleaning sheet no.3 ***Total No. of washing of modules in current month***

Date	**Section of the plant**	**Section of the plant**	**Section of the plant**	**Remark**

Daily plant fault and outage monitoring sheet no.4

	Plant name………………….. Plant Capacity…………………………Month………………..
Date	

Daily grid fault and outage report sheet no 5.

	Name of the plant……………… Capacity of the plant…………………..	
Date	Grid Fault/Outage Time	Reason of grid fault as per GSS report
	–	
	–	
	–	
	–	
	–	
	–	

	–	
	–	
	–	
	–	
	–	
	–	
	–	
	–	
	–	
	–	
	–	
	–	
	–	

O&M Monthly Log Sheet for -----2014 for the plant…………Capacity……..MW

SUMMARY OF THE TOTAL MONTHLY GENERATION OF --------2014								
MONTHLY	Month till date generation for -----2014							
	Month till date generation including Aux supply for -------2014							
	Month till date Generation Target for ---------2014							
	Month till Date Generation loss bcz of grid failure for ------- 2014							
	Month till Date target excluding grid loss for --------2014							
	% Gen Target Achieved MTD							
WEEKLY	Week till date Generation for Week 4							
	Week till date Generation including AUX supply for Week 4							
	Week till date Generation Target for Week 4							
	Week till Date Generation loss bcz of grid failure for Week 4							

	Week till Date target excluding grid loss for Week 4									
	% Gen Target Achieved WTD									
Daily Avg global solar radiation (sunshine hours)Wh/m2	Total daily power generation (KWh)	Net Exportable Power (KWh)TVM Four pole	Weather	Number of grid failures and total grid outage		Grid Availability	Generation Loss due to Grid	Daily Generation Target	Daily Generation Target excluding grid loss	
				Nos.	No of hrs/day					

Made in the USA
Monee, IL
08 November 2026

67972895R00070